Parasitic-Aware Optimization of CMOS RF Circuits

by
David J. Allstot
Kiyong Choi
Jinho Park
University of Washington

Kluwer Academic Publishers
Boston/Dordrecht/London

Distributors for North, Central and South America:
Kluwer Academic Publishers
101 Philip Drive
Assinippi Park
Norwell, Massachusetts 02061 USA
Telephone (781) 871-6600
Fax (781) 871-6528
E-Mail <kluwer@wkap.com>

Distributors for all other countries:
Kluwer Academic Publishers Group
Post Office Box 322
3300 AH Dordrecht, THE NETHERLANDS
Telephone 31 78 6576 000
Fax 31 78 6576 474
E-Mail <orderdept@wkap.nl>

 Electronic Services <http://www.wkap.nl>

Library of Congress Cataloging-in-Publication

CIP info or:

Title: Parasitic-Aware Optimization of CMOS RF Circuits
Author (s): David J. Allstot, Kiyong Choi & Jinho Park
ISBN: 1-4020-7399-2

Dedication

This book is dedicated to
Sarah and Matthew,
To our Lord and Hyun-Ah,
and to Vickie, Kevin, and Emily

Contents

Contributing Authors

Kiyong Choi
Jinho Park
David J. Allstot

Preface

The annual market for wireless devices exceeds tens of billions of dollars worldwide. As markets expand and evolve, there is an insatiable demand for greater functionality in smaller form factor devices, seamless compatibility with various communications standards, longer battery operating lifetimes, and, of course, lower costs. The confluence of these objectives has motivated worldwide research on system-on-chip (SOC) or system-in-package (SIP) solutions wherein the number of off-chip components is relentlessly driven towards zero. These objectives have in turn motivated the development of CMOS and BiCMOS technologies that are effective in implementing digital, analog, radio frequency, and micro-electro-mechanical functions together in SOC solutions.

In the arena of RF integrated circuit design, efforts are aimed at the realization of true single-chip radios with few, if any, off-chip components. Ironically, the on-chip passive components required for RF integration pose more serious challenges to SOC integration than the active CMOS and BJT devices. Perhaps this is not surprising since modern digital IC designs are dominated as much, or more, by interconnect characteristics than by active device properties. In any event, the co-integration of active and passive devices in RFIC design represents a serious design problem and an even more daunting manufacturing challenge. If conventional mixed-signal design techniques are employed, parasitics associated with passive elements (resistors, capacitors, inductors, transformers, pads, etc.) and the package effectively de-tune RF circuits rendering them sub-optimal or virtually useless. Hence, dealing with parasitics in an effective way as *part of the design process* is an essential emerging methodology in modern SOC design. The parasitic-aware RF circuit synthesis techniques described in this book effectively address this critical problem.

In conventional mixed-signal design, parasitic resistances and capacitances are e stimated d uring t he d esign p rocess a s they a ffect d esign parameters such as bandwidth, phase margin, slew rate, etc. Circuits such as opamps are characterized by one-sided requirements on the parasitics. That is, so long as a parasitic capacitance is below a certain value, for example, the phase margin specification is met or exceed. Because of these one-side requirements, the circuits are usually designed making some initial assumptions about the parasitics and optimized just once manually based on the parasitics extracted from the layout.

Traditional mixed-signal design practices simply do not work for RF circuits because of the two-side requirements on their parasitic element values. To tune an RF amplifier to a specific frequency, for example, a nodal capacitance can be neither too large nor too small. That is, it must realize a certain value within a specified tight tolerance. This means that the parasitics associated with all passive elements must be accurately modeled as a function of the element value prior to the synthesis process. For on-chip spiral inductors, for example, the compact model often includes about 10 parasitic components that are complex functions of frequency and inductance value. Moreover, since typical RF circuit blocks may employ tens of passive components, more than 100 passive and parasitic element values must be included in the design process. Obviously, such complexity is unwieldy for typical hand calculations scribbled on the back of a cocktail napkin. Hence, fast and aggressive optimization tools must be readily available to assist the designer in meeting specifications with robustness and cost-effectiveness.

Chapter 1 introduces basic wireless RF transceiver circuits that are amenable to the parastic-aware circuit synthesis paradigm. Chapter 2 provides an overview of the various passive components used in RF circuit design and techniques for creating compact parametric models that are computationally efficient when used in the parasitic-aware optimization loop. Possible optimization strategies for parasitic-aware synthesis are presented in Chapter 3 including the classical gradient decent, simulated annealing, and genetic algorithms. Since parasitic-aware synthesis currently requires hours or days of computer time even for relatively simple circuits, efficiency in the optimization algorithms in reducing the required number of interations is of paramount importance. Hence, new optimization strategies including t unneling a nd a daptive t emperature coefficient h euristics f or u se with simulated annealing are presented along with an exciting and interesting particle-swarm methodology that finds its first use in circuit synthesis in this book. Part II of the book provides background on many of the key RF circuit blocks and focuses on their parasitic-aware synthesis. Specifically, Chapters 4 through 8 cover low-noise amplifiers, up- and down-conversion mixers, voltage-controlled oscillators, power amplifiers, and wideband distributed amplifiers, respectively.

We first developed the proposed parasitic-aware design methodology in 1995 in conjunction with Brian Ballweber and Dr. Ravi Gupta. Since then, our students and colleagues have improved the design paradigm greatly and applied it to many interesting and exciting circuits. This book reflects a small part of that experience. We are especially thankful to Adam Chu for writing most of Chapter 2, and to our current graduate students for their contributions: Sankaran Aniruddhan, Sherjiun Fang, Taeik Kim, Srinivas Kodali, Waisiu Law, Seetaur Lee, Xiaoyong Li, Kristen Naegle, Dicle Ozis, Jeyanandh Paramesh, Charles Peach, Brian Ward, and Hossein Zarei.

We are pleased to acknowledge support for the research work described herein from the following sources: DARPA NeoCAD Program under grant N66001-01-8919; Semiconductor Research Corporation grants 2000-HJ-771 and 2001-HJ-926; National Science Foundation contracts CCR-0086032 and CCR-01200255 and MRI-0116281, and grants from the National Science Foundation Center for the Design of Analog-Digital Integrated Circuits, Texas Instruments, National Semiconductor, and Intel Corporation.

PART I: BACKGROUND ON PARASITIC-AWARE OPTIMIZATION

Chapter 1

INTRODUCTION

1.   INTRODUCTION

In the near future, people's daily activities will be dominated with portable wireless devices. To make compact and energy efficient mobile communicating equipment such as mobile phones, wireless modems, two-way radios, etc., integrated circuit technology (IC) is necessary because it not only reduces the size, weight, and power consumption of such devices, it enables manufacturers to include greater functionality at lower costs. Figure 1-1 shows a general system-on-chip (SOC) solution that comprises digital, analog, micro-electro-mechanical systems (MEMS), and radio frequency (RF) circuit blocks. By implementing a sensor in conjunction with an RF front end, the SOC can receive or transmit data from external sources. The digital circuitry on the chip consists of CPU, DSP, and memory blocks, and is used for data processing. Analog-to-digital converters (ADC) and digital-to-analog converters (DAC) enable the communication of information between the digital and analog circuits.

It was not long ago that all RF circuits were implemented using gallium arsenide (GaAs) or bipolar junction transistor (BJT) technologies. CMOS technology was not viable due to its low values of breakdown voltage, small-signal transconductance, and unity current gain frequency f_t. Moreover, the lossy silicon substrates contributed a plethora of parasitic elements to the monolithic passive components that made CMOS inferior to its bipolar and GaAs counterparts.

However, because CMOS has dominated the world of digital ICs for more than a quarter century and has achieved a very high level of integration, it has become an extremely compelling and cost effective option for use in SOC design [1],[2]. It has been shown that mixed-signal functions such as ADCs are effectively designed in CMOS [3],[4]. But despite a large number of books and other publications [5]-[7], critical RF circuitry such as low-noise amplifiers (LNA), up- and down-conversion mixers, voltage-controlled oscillators (VCO), power amplifiers (PA), and wide-band amplifiers still pose difficult challenges to circuit designers. Alternative solutions such as multi-chip modules and System-in-Package (SIP) solutions along with the use of the traditional GaAs technology are difficult to implement or expensive. Thus, in order to exploit CMOS as the technology solution for high-volume SOC design and manufacture, the inferior nature of the technology, especially with respect to parasitic elements, needs to be overcome. Our experience over the past several years has shown that one way to achieve high performance designs with robust manufacturing characteristics vis-a-vis process, temperature, and voltage variations is to perform global optimization of CMOS RF integrated circuits by including all device and package parasitics as *part of the design process*.

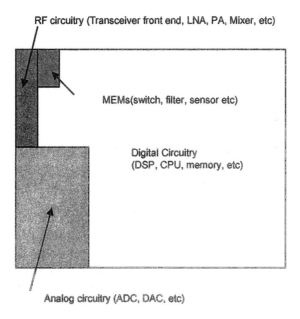

Figure 1-1 System-on-chip solution

2. OVERVIEW OF WIRELESS TRANSCEIVERS

When it comes to wireless communications, there are many system standards to support myriad applications. For example, both analog and digital modulation schemes have been implemented at various RF frequencies. Amplitude modulation (AM) and frequency modulation (FM) are examples of analog modulation methods. Digital modulation techniques are categorized as either linear or nonlinear. Binary Phase-Shift Keying (BPSK), Quadrature Phase-Shift Keying (QPSK), Amplitude Shift Keying (ASK), 16-ary Quadrature Amplitude Modulation (16QAM), 9-state Quadrature Partial Response (9QPR), etc., are examples of the former, while Minimum Shift Keying (MSK), Gaussian Minimum Shift Keying (GMSK) and Frequency Shift Keying (FSK) are examples of the latter. The wide variety of standards also imposes a number of RF front-end circuit requirements. For example, if the modulation frequency of the system is 900 MHz, the front-end circuits need to be tuned through the use of passive components to 900 MHz; if the system uses a linear digital modulation scheme, a highly linear front-end RF circuit is required, and so on.

Figure 1-2 shows a basic RF front-end homodyne transceiver that operates directly between RF and baseband frequencies. On-chip implementations include low-noise amplifier (LNA), mixer, and power amplifier (PA) building blocks; critical off-chip components usually include a duplexer, and various band-select, image-reject, and channel-select filters.

The off-chip duplexer is used to switch the external antenna between transmit and receive paths of the transceiver. Important specifications of the duplexer are isolation between transmit and receive paths as well as its switching loss which adds directly to the overall noise performance of the transceiver system.

The main purpose of the LNA in the receive path is to provide sufficient gain for the RF signal to overcome the noise of subsequent stages of the receiver. In addition, its gain must be low enough to accommodate the very wide dynamic range of the input signal without saturating the LNA. Usually, the noise of the LNA dominates the noise performance of the on-chip portion of the receiver system.

The down-conversion mixer in the receive path of the homodyne receiver of Fig. 1.2 is used to directly translate the RF input signal to the baseband frequency, while the up-conversion mixer in the transmit path performs the opposite function. In heterodyne architectures, a first down-conversion mixer in the receive channel converts the RF frequency to an intermediate frequency (IF) and a second down-conversion mixer converts from IF to baseband, while two up-conversion mixers perform the opposite functions in the transmit channel.

A special type of mixer, called a quadrature mixer, creates both in-phase (I) and quadrature (Q) IF signals whose phases ideally differ by exactly 90 degrees. One of the mixer inputs is from the local oscillator (LO) signal that is usually taken from a VCO that is embedded in a frequency synthesizer. The general requirements on mixers are that they can be relatively noisy compared to LNA circuits, for example, but they must exhibit relatively high linearity.

Several off-chip band pass filters are used to limit the frequency ranges at various points in the transceiver. Band-select band pass filters are often used at the input to the LNA in the receive path and at the output of the PA in the transmit path. An image rejection filter (IRF) is used after the LNA at the input to the receive down-conversion mixer; likewise, a band pass filter is inserted at the output of the mixer to select the desired IF signal from among several undesired modulated frequencies.

A power amplifier (PA) is a challenging component of transmitter front-end used to amplify the RF signal to very high levels for transmission from the external antenna. In fact, the output power of the PA is determined by the desired maximum transmit range. Since the PA is a major source of power dissipation in the transceiver, its efficiency is very important in determining the overall efficiency of transceiver and its battery life. There are two major types of power amplifiers; namely, linear and nonlinear. As mentioned earlier, the type of PA required in a transmit channel of a system is determined by its modulation scheme. For example, QPSK requires a linear PA while GMSK can take advantage of a nonlinear PA. Usually, the power efficiency of a linear PA is significantly lower than that of its nonlinear brethren. The power amplifier presented in Chapter 7 of this book is a high efficiency nonlinear class-E PA for GMSK applications.

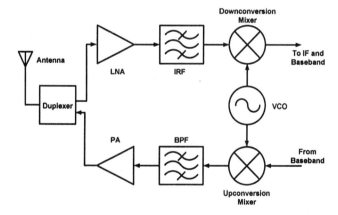

Figure 1-2 Block diagram of typical homodyne RF transceiver

3. OUTLINE OF THE BOOK

The primary objective of this book is to study optimization methods as applied to the major circuit blocks in RF IC design. The optimization techniques described apply not only to the circuits in this book, but to virtually all circuits as well as applications in other fields.

This book is divided into two parts. Part I discusses the background on parasitic-aware optimization; Chapter 2 provides an overview of the modeling of monolithic passive and active components for use in parasitic-aware synthesis. Part II of the book introduces the readers to optimization techniques for various RF circuits. Specifically, Chapter 3 discusses the general concepts as well as algorithms for optimization, while Chapters 4, 5, 6, 7, and 8 investigate the parasitic-aware optimization of LNA, MIXER, VCO, PA, and wide-band amplifier RF circuits blocks, respectively.

REFERENCES

[1] G. E. Moore "Cramming m ore components onto integrated circuits," *Electronics*, Vol. 38, Number 8, April 1965.

[2] http://www.semichips.org, *National Roadmap for Semiconductors*, Semiconductor Industry Association, 2002.

[3] P.R. Gray, B.A. Wooley, and R.W. Brodersen, "Analog MOS Integrated Circuits, II," IEEE Press, 1998.

[4] S.R. Norsworthy, R. Schreier, and G.C. Temes, *Delta Sigma Data Converters: Theory, Design, and Simulation*, IEEE Press, 1996.

[5] B. Razavi, *RF Microelectronics*, Prentice Hall, New Jersey, 1998.

[6] T.H. Lee, *The design of CMOS Radio-Frequency Integrated c ircuits*, Cambridge University Press, 1998.

[7] B. Leung, *VLSI for Wireless Communication,* Prentice Hall, 2002.

Chapter 2

MODELING OF ON-CHIP PASSIVE AND ACTIVE COMPONENTS

Unlike traditional analog IC design that often employs capacitor and resistor passives, RF IC design also makes significant use of on-chip inductors and transformers. The phase noise and the tuning range of voltage-controlled oscillators and the matching requirements for low-noise amplifiers and power amplifiers are just a few examples of how design specifications can be dictated by the quality of such on-chip passive components. Needless to say, the design and layout of monolithic passive components is extremely important in mixed-signal and RF designs. At the same time, higher-order effects in active devices that are usually not present and thus not considered in base-band circuit designs often dominate. This chapter provides an overview of on-chip passive and active components and at the same time introduces an inexpensive yet effective method for parasitic-aware inductor modeling. The material in this chapter is intended as a brief overview rather than a rigorous review of the subject.

1. MONOLITHIC INDUCTORS

1.1 Background on monolithic inductors

Traditionally, RF IC designs have incorporated off-chip surface-mounted inductors. While these inductors offer extremely high quality factor (Q) and generally good performance, they are detriments to board-level-complexity and overall system cost reduction. An alternative to off-chip inductors is bonding wires, but their main shortcoming in a manufacturing environment is the lack of tight control of the bonding process [1]. This issue is addressed in a later section.

Monolithic spiral inductors have traditionally been fabricated on GaAs substrates because in addition to providing a substrate that is a very good insulator, the process often offers an electroplated gold top-most interconnect layer that has high conductivity. The inductor Q associated with typical GaAs processes ranges from ten to twenty. Silicon technologies, on the other hand, use low substrate resistivities ranging from 10Ω-cm to 100Ω-cm for BICMOS processes, and less than 1Ω-cm substrates for standard CMOS processes. At the same time, the metal interconnect layers are usually aluminum-based with relatively low conductivity values. Typical inductor Q values associated with silicon processes are less than ten, sometimes as low as three or four. These characteristics present a dilemma to circuit designers: Cost reduction calls for the use of silicon processes while the increasingly stringent design specifications suggest an expensive GaAs technology [1].

1.2   Monolithic inductor realizations

The inductance and other characteristics of a monolithic spiral inductor are primarily determined by *process-controlled* parameters including [2]

- Oxide thickness
- Substrate resistivity
- Number of metal layers
- Thickness and composition of metal layers (e.g., Al, Au, or Cu)

as well as its *design-controlled* lateral parameters such as [2]

- Number of turns, n
- Metal width, w
- Edge-to-edge spacing between adjacent metal, s
- Outer or inner diameter, d_{out} or d_{in}
- Number of sides, N

Sometimes, the average diameter or the fill ratio, defined as $d_{avg} = d_{in} + d_{out}$ and $\rho = (d_{out} - d_{in})/(d_{out} + d_{in})$, respectively, are used instead of the inner or outer diameters. The parasitic element values are determined by the lateral and vertical parameters of the spiral inductor. Usually, top metal is used because its large thickness reduces the DC series resistance and its distance from the substrate reduces the metal-to-substrate capacitance. In most RF oriented processes, "bump" or RF metal interconnect layers are available for this purpose. Figure 2-1 shows the cross-sectional and top views of a square spiral inductor.

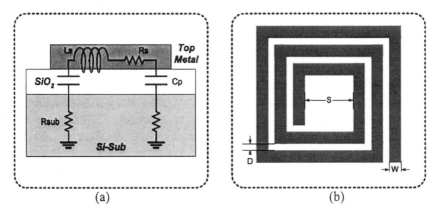

          (a)   (b)

Figure 2-1 Parasitic-laden monolithic square spiral; (a) cross-sectional view and (b) top view

Spiral inductors can take on different geometries. The most often implemented is the square or "Manhattan" spiral due to its layout ease and support from all layout tools. The most optimum geometry, however, is the circular spiral because it places the largest amount of metal in the smallest amount of area, thus greatly reducing the loss resistance and capacitance [2]. Since very few commercial CAD layout tools support the circle-drawing feature, it is not very common to see circular spiral inductors in SOC solutions. Hexagonal and octagonal structures, which approximate the circular structure to a certain degree, have been used instead since they are easier to layout and most CAD tools can accommodate 45° angles.

1.3   Monolithic inductor models

Traditionally, inductor model extraction techniques are categorized into three different groups [2]:
1. EM solvers
2. Distributed/segmented models
3. Lumped circuit models

The EM solvers use the most accurate approach to model inductors, which is the numerical solving of Maxwell's equations. By applying boundary conditions, tools such as *SONNET*[1] and *ADS*[2] can solve Maxwell's

---

[1] SONNET is a registered trade-mark of Sonnet Software, Inc.

equations numerically. Even though very accurate, these simulators are not suitable for simulating monolithic spiral inductors because simulation can take days or even weeks to complete. Also, besides mandating access to detailed process information of the technology that is usually not available, these tools have a fairly steep learning curve and require users to be quite experienced in their use.

In order to reduce simulation time and increase efficiency, custom simulators have been developed just for the purpose of simulating spiral inductors. One such tool that is very well known is *ASITIC*[3] developed by Niknejad [3]. It achieves fast simulation speed compared to the tools mentioned above by using electrostatic and magnetostatic approximations to solve matrices. Its major shortcomings on the other hand, are the lack of an interface between it and standard circuit simulators such as *SPICE*[4] and *SPECTRE*[5], as well as the lack of design insight it provides on engineering trade-offs to circuit designers, which is, of course, characteristic of all CAD tools.

The second approach models each segment of the spiral as an individual transmission line network, thus forming a distributed model for the entire inductor. The distributed model of a single-turn square spiral inductor is shown in Fig. 2-2. The parasitic resistances and capacitances are determined using the Greenhouse approach [4]. Even though this method is more efficient than the EM solver, the shortcoming is still the size and complexity of the resulting model. The size of a multiple-turn polygon structure can easily surpass that of the rest of the RF circuit, greatly increasing simulation time. Optimization is also difficult to perform since it requires a program that dynamically changes the number of segments in the model [2].

The last method, which is preferred by most circuit designers, is the lumped model wherein the spiral is approximated by an equivalent π-network with inductance in series with the loss resistance and parasitics shunted to ground as shown in Fig. 2-3. The major advantage of this model is of course speed with the major tradeoff being accuracy. It turns out that as long as the frequency of operation is well below the resonant frequency of the spiral, the lumped model provides sufficient accuracy [2],[5].

---

[2] ADS—Advanced Design System—is a registered trade-mark of Agilent Technologies
[3] ASITIC— Analysis and Simulation of Spiral Inductors and Transformers for ICs— http://formosa.eecs.berkeley.edu/~niknejad/asitic.html
[4] SPICE— Simulation Program with Integrated Circuit Emphasis
[5] SPECTRE— is a registered trade-mark of Cadence

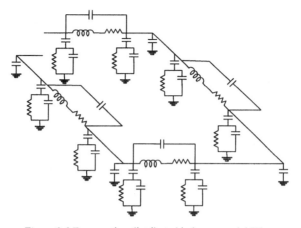

Figure 2-2 Four-section distributed inductor model [2]

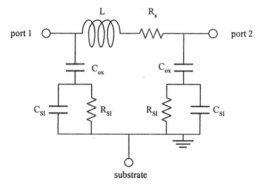

Figure 2-3 Lumped compact model for a monolithic inductor

1.4 Expressions for the lumped inductor model

An analytical expression for the lumped compact model shown in Fig. 2-3 is presented in [2] and shown below:

$$L = \sum_{i=1}^{n} L_i + \sum_{i=1}^{n} \sum_{j=1, j \neq i}^{n} M_{i,j} \approx 6 \times 10^{-7} n^2 d_{out} \tag{2-1}$$

where the first summation term indicates the self-inductances of the segments and the second summation term represents all mutual inductances.

A rough approximation for the inductance value of a square spiral with an error that is less than 20% is shown in the above equation as well [6]. The inductance computation is done using the Greenhouse method, which states that the inductance of a spiral is the sum of the self-inductance of each segment and the positive and negative mutual inductance between all possible segment pairs [5]. The mutual coupling signs are determined by the directions of currents in the segments: + if same, - if different; no mutual inductance exists between two segments that are orthogonal. The DC series resistance of the spiral is given by Eq. (2-2):

$$R_s \approx \frac{l}{\sigma w \delta \cdot \left(1 - e^{-t/\delta}\right)} \qquad (2\text{-}2)$$

where σ is the conductivity, δ is the skin depth, ω is the metal width, μ is the magnetic permeability and l is the length of the spiral, which is approximated as $l = n \cdot d_{avg} \cdot N \cdot tan(\pi/N)$.

The metal-to-substrate oxide capacitance C_{ox} is the major contributor to parasitic capacitance of the spiral inductor. An expression for evaluating it is written as [2]:

$$C_{ox} \approx \frac{1}{2} \frac{\varepsilon_{ox}}{t_{ox}} l w \qquad (2\text{-}3)$$

where ε_{ox} is the oxide permittivity that has a value of 3.45×10^{-13}F/cm and t_{ox} is the oxide thickness between the spiral and the substrate.

The feed-through (series) capacitance C_F models the capacitive coupling between the input and output ports of the spiral. Its contribution comes from the overlap between the spiral itself and the underpass and the crosstalk between adjacent turns. In most cases, the crosstalk capacitance is negligible and can be ignored, so the feed-through capacitance is modeled by the following equation [5]:

$$C_F \approx \frac{\varepsilon_{ox}}{t_{ox,ind-underpass}} nw^2 \qquad (2\text{-}4)$$

where n is the number of overlaps, w is the metal width, and $t_{ox, ind-underpass}$ is the oxide thickness between the spiral and the underpass.

The substrate capacitance and resistance can be expressed as [2]:

$$C_{si} \approx \frac{1}{2} C_{sub} l w \qquad (2\text{-}5a)$$

$$R_{si} \approx \frac{2}{G_{sub} l w} \qquad (2\text{-}5b)$$

where C_{sub} and G_{sub} are the substrate capacitance and conductance per unit area, respectively. The factor of 2 in Eq. (2-5a) and Eq. (2-5b) account for the fact that the parasitics are distributed equally at the two ends of the spiral. Both C_{sub} and G_{sub} are generally functions of doping and can be easily extracted from measurements.

1.5 Monolithic transformers

Traditionally, monolithic transformers have been used in microwave and RF ICs for matching and phase splitting purposes. Recently however, they have seen applications in LNA and VCO designs due to their smaller area and effective higher Q values [7],[8]. In this section, the characterization of monolithic transformers is presented.

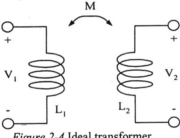

Figure 2-4 Ideal transformer

An ideal transformer, shown in Fig. 2-4, can be expressed as follows [2]:

$$v_1 = L_1 \frac{\partial i_1}{\partial t} + M \frac{\partial i_2}{\partial t} \qquad (2\text{-}6a)$$

$$v_2 = L_2 \frac{\partial i_2}{\partial t} + M \frac{\partial i_1}{\partial t} \qquad (2\text{-}6b)$$

where L_1 and L_2 are the self inductances of the primary and secondary windings, respectively, and M is the mutual inductance between the primary and the secondary. The coupling coefficient k that relates the two self inductances, is defined as:

$$k = \frac{M}{\sqrt{L_1 L_2}}$$ (2-7)

For an ideal transformer, $k = 1$. In practical on-chip transformers, k typically takes on a value between 0.2 and 0.9, depending on the realization [9]. The three most commonly realized transformer structures are the concentric structure, the interleaved structure, and the stacked structure.

The concentric structure, shown in Fig. 2-5, relies on lateral magnetic coupling and is asymmetric. The spirals are implemented with top level metal to reduce series resistance. Since the common periphery between the two windings is just a single turn, capacitive coupling is minimized; unfortunately, mutual inductance, and thus the coupling coefficient are greatly reduced for the same reason. The k factor for this structure is typically below 0.5 [2]. Even though the low coupling coefficient limits the applications of the concentric structure, it can be useful in applications that require low mutual-to-self inductance ratio, such as broadband amplifiers [9].

The interleaved structure, shown in Fig. 2-6, is a symmetrical structure wherein the coupling is achieved by alternating the primary and secondary windings as shown. The electrical characteristics of the two windings are the same when they have the same number of turns. A major advantage of this structure is that the terminals of the windings are designed to be on opposite sides, which facilitates the wiring between it and other circuits. The k for this structure is usually around 0.6. Higher coupling coefficients can be achieved by decreasing the metal width, but at the cost of higher series resistance [2],[9].

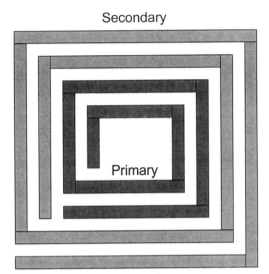

Figure 2-5 Concentric structure

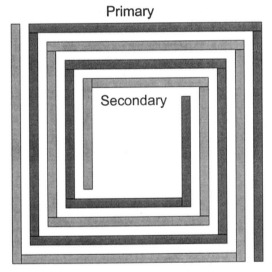

Figure 2-6 Interleaved structure

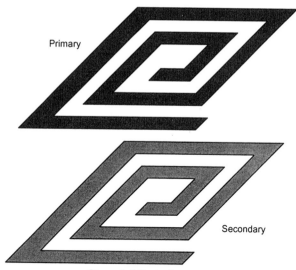

Figure 2-7 Stacked structure

The stacked structure, shown in Fig. 2-7, uses two different metal layers to achieve both lateral and vertical magnetic coupling. This structure provides good area efficiency and high coupling coefficient, usually above 0.8. One shortcoming of this configuration is that the electrical responses of the primary and secondary are different even though they are geometrically identical. This difference mainly arises from the difference in thickness between metal layers as well as the different metal-to-substrate capacitances [9]. A second drawback is that the large metal-to-metal capacitance leads directly to a low self-resonant frequency. In order to alleviate this problem, in a multi-metal process, instead of using the top two metal layers, the top and the third from the top metal layers are used to increase the distance between the spirals. According to [2], this technique can reduce the capacitance by more than 50% while only reducing the coupling coefficient by 5%.

1.6   Monolithic 3-D structures

The stacked inductor introduced in the previous section is an example of an on-chip 3-D structure. In the past decade, low Q and self-resonant frequency (f_{sr}) values, and the large area requirements of single-layer inductors have motivated a significant research effort. Innovative spiral geometries such as the wind-in wind-out inductors and their derivatives have been reported to achieve higher Q and f_{sr} [8],[10].

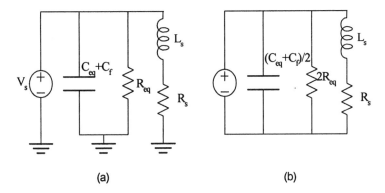

Figure 2-8 Equivalent lumped inductor model (a) single-ended configuration (b) differential configuration

A subtle yet important issue in inductor design is symmetry. Two simplified models for a traditional square spiral are shown in Fig. 2-8 (a) and 2-8(b), respectively. The difference between the two is that the model in Fig. 2-8(a) is connected in a single-ended configuration, while the one in Fig. 2-8(b) is connected differentially. The models show a parasitic capacitive branch with capacitors C_f and C_{eq} and resistance R_{eq} in parallel with the inductive branch. The capacitive branch has high impedance when the inductor is excited differentially. Hence, the differentially excited structures are inductive over a larger range of frequencies and therefore have higher Q and f_{sr} values. This improvement in performance is inherent to all differentially excited structures. In the asymmetric structures, however, the parasitic impedances looking into the two ports take on two different values and hence, the identification of a common-mode point is difficult which restricts the use of these structures in differential applications. One solution to this problem is to use two asymmetric inductors in series [11].

Recently, a fully symmetric 3-D inductor was proposed in [11] and is shown in Fig. 2-9. A combination of the wind-down wind-up and the wind-in wind-out structures, the proposed spiral winds in and down alternately after every half turn. This structure provides positive mutual coupling and avoids the need of having two separate inductors. The port locations of this structure can also be advantageous in circuit layout. Simulation shows that this structure consumes 58% less area compared to that of a planar spiral, and provides a higher f_{sr}. However, the Q is slightly lower than that of the planar spiral because it necessarily uses thinner lower metal layers with many resistive vias between layers. Fig. 2-10 shows the performance characteristics of the structure.

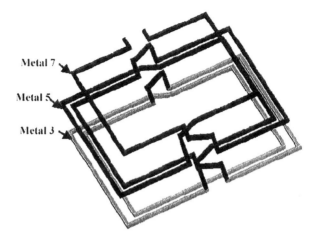

Figure 2-9 Fully symmetric 3-D miniature inductor in [11]

1.7　Parasitic-aware inductor model

The three modeling methods mentioned in Section 2.4 all have their merits as well as shortcomings. An alternative way to model spiral inductors is to fabricate several spiral inductors and calibrate an EM simulator based on the measurement results. After calibration, the EM simulator is accurate in calculating parasitics for interpolated inductance values. The key merit for this method is that it is very accurate because no assumptions are made with measurements of fabricated inductors, and also, it is relatively inexpensive since it doesn't require fabrication of every inductor needed for a design. Usually only three inductors with identical metals widths, spacings, etc., need to be fabricated and tested, and then arbitrary desired inductance values can be interpolated.

Figure 2-11 shows a typical measured frequency response of an on-chip spiral inductor. In 0.35µm CMOS, a metal-3 spiral with 6.25 turns (9nH, metal width=15µm, center spacing=101.4µm, and metal spacing =1.2µm) exhibits a peak Q of 4 at 3.2GHz with a self-resonant frequency of 3.2GHz. Parasitic-aware modeling requires that the parasitic values be expressed in terms of the design variables. Figure 2-12 shows such relationships versus inductance values obtained using the *polyfit* command in *MATLAB*[6]. To obtain the relationships of the parasitics in terms of inductance, three different inductors are first fabricated and measured. Based on the

---

[6] MATLAB is a registered trade-mark of Mathworks

measurement results, other points are calculated using an EM simulator such as *ADS*. Then, using the *polyfit* command in *MATLAB*, equations like those shown in Fig. 2-12 are developed.

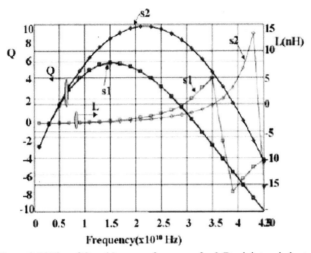

Figure 2-10 Plot of Q and L versus frequency for 3-D miniature inductor

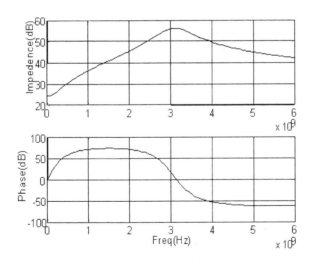

Figure 2-11 Typical measured frequency response of an on-chip square spiral inductor

Another possible on-chip inductor is the bond wire, which has several advantages over the spiral inductor. Since the composition of the bonding wire is usually gold rather than aluminum, and its radius is large (about 30μm) compared to the dimensions of the spiral, its series resistance is much smaller. This means that its quality factor is much higher than that of an on-chip spiral. Figure 2-13 shows a bond wire inductor with its important parasitics. The bonding pads contribute additional C_{ox} and R_{sub} terms that do not depend on the inductance.

As a rule of thumb, each 1mm length of the bond wire contributes 1nH of inductance. A first-order expression to calculate the inductance value is [12]:

$$L = \frac{l}{5} \cdot (\ln(2 \cdot \frac{l}{r}) - 0.75 + \frac{r}{l})$$
(2-8)

where L is inductance in nH, l is the length of the bond-wire in mm, and r is its radius in mm. Typical values of the pad capacitance and resistance are 200 aF for C_{pad} and 20Ω for R_{pad}.

The series resistance is given by Eq. (2-2). At 900 MHz, the skin depth of gold is 2.57 μm. The radius of the bond-wire inductor is typically 30 μm. The area of the conduction part can be approximated as:

$$A_{eff} = 2 \cdot \pi \cdot r \cdot \delta$$
(2-9)

Using Eqs. (2-2), (2-8), and (2-9), the series resistance is calculated as 0.05Ω/mm at 900MHz.

Manufacturability and repeatability of bond wire inductors have previously been concerns. However, it has recently been shown that machine-bonded wires of the type used in manufacturing exhibit less than ±5% inductance variations and less than ±6% Q variations [13].

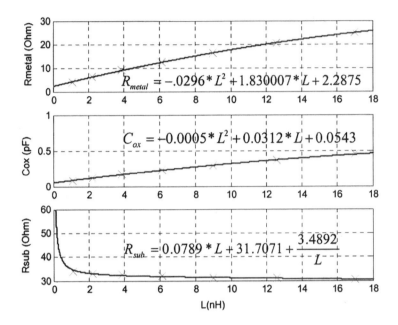

$$R_{metal} = -.0296 * L^2 + 1.830007 * L + 2.2875$$

$$C_{ox} = -0.0005 * L^2 + 0.0312 * L + 0.0543$$

$$R_{sub} = 0.0789 * L + 31.7071 + \frac{3.4892}{L}$$

Figure 2-12 Parasitics versus inductance for on-chip square spiral inductors

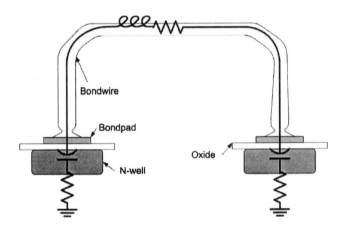

Figure 2-13 A bond wire inductor with parasitics

2.   MONOLITHIC VARACTORS

When it comes to RF IC design, on-chip varactors have usually been taking a back seat compared to their inductive brethrens. The reason is that the quality factor (Q) of the tank is often limited by the Q of the inductor. But as the Q of the inductor continues to rise due to technology improvements, varactor design has become just as important.

Traditionally, two types of varactors are available to RF designers: diode varactors, and MOS varactors. The subsequent sections examine the physical operation as well as the merits and shortcomings of each type.

2.1   Diode varactor

The pn-junction diode without question is the most fundamental form of varactor in integrated circuits. A diagram of a typical abrupt pn-junction is shown in Fig. 2-14. The existence of an effective capacitance, which is created by the majority carriers in the junction transit region, can be expressed in the form of a parallel plate capacitor. A schematic of the physical diode is shown in Fig. 2-15 to illustrate the capacitance associated with the depletion region. As a reverse-biased voltage is applied across the junction, the width of the depletion region increases and the capacitance decreases. The junction capacitance C_j is given as [14]:

$$C_j = \frac{C_{j0}}{\left[\left(V_R / \phi_0\right)+1\right]^{\alpha_j}}$$

(2-10)

where C_{j0} is the zero-biased capacitance value, ϕ_0 is the built-in potential, V_R is the reverse-biased voltage, and α_j is the junction profile coefficient. It is also very important to note that the width of the depletion region and thus the capacitance value depends on the junction profile. In the case of an abrupt junction, the charge density changes almost linearly with respect to the reverse-biased voltage ($\alpha_j=1/2$); in the case of a linearly graded junction, the charge density change increases with additional increment in applied voltage ($\alpha_j=1/3$).

In conjunction with the capacitance produced are losses that deviate diode behavior from that of a perfect capacitor. Several sources contribute to the losses in a diode, but the most dominant one is the n-type base resistance, shown in Fig. 2-15. In order for the capacitance to change when a voltage is applied, the doping level of this n-type region needs to be low. This directly translates to low conductivity and thus high resistance. On the other hand, if a highly doped base region is applied, the depletion region capacitance is virtually constant. The typical base resistance is on the order of an ohm. The second most important source of losses is the series resistance. The series resistance is usually defined as the sum of R_{ns} and R_{ps}, where both R_{ns} and R_{ps} include resistance of the N$^+$ and P$^+$ regions, respectively, as well as contact resistances. To make matters worse, the series resistance exhibits skin effect at high frequencies. This can be on the order of several hundred milliohms or higher and introduces significant loss for high-Q varactors. The third source of loss is the bulk and surface leakage around the depletion layer, represented by $R_j(V)$ in Fig. 2-15. At high frequencies, this loss is negligible since it is shunted by the depletion region capacitance. But at low frequencies, this loss mechanism can be significant and dictates the Q of the varactor [15].

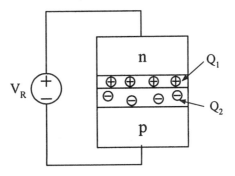

Figure 2-14 Reverse-biased abrupt pn-junction

The shortcomings of diode varactors are smaller tuning ability and higher noise performance due to their lower Q values compared to MOS varactors, nonlinear C-V behavior, and potential problems with forward-biasing the p-n junction if the signal swing is large. Also, a standard digital CMOS process may not have the device model for it since the diode is typically not used in digital IC design.

2.2   Inversion-mode MOS varactors

The MOS capacitor, sometimes referred to as the two-terminal MOSFET, can be formed by connecting the drain, source and substrate, and is shown in Fig. 2-16 [14]. Before discussing the behavior of such a capacitor, which depends on the gate-to-substrate voltage (V_{GB}), it is necessary to review the definition of the flat-band voltage because it dictates the region the MOS operates in. Consider a case in Fig. 2-17, where the gate is made of material

A and the substrate is made of material B. It is shown in [14] that if the gate and the substrate are shorted together, the contact potential is given by

$$\phi_{contact} = \phi_A - \phi_B \qquad\qquad (2\text{-}11)$$

where ϕ_A and ϕ_B are the contact potentials of the gate and substrate, respectively.

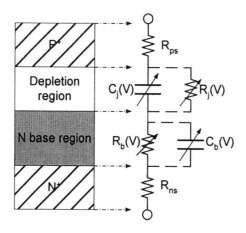

Figure 2-15 Physical and electrical schematics of a diode varactor

Because of this non-zero potential between the gate and the substrate, charges will appear on both sides of the oxide. If an external potential ϕ_{ms} is applied that cancels the contact potential in Eq. (3-1), then ϕ_{ms} is called the flat-band voltage. From Eq. (3-1), ϕ_{ms} must have a value given by:

$$\phi_{ms} = \phi_B - \phi_A \qquad\qquad (2\text{-}12)$$

With the flat-band voltage V_{FB} defined, the capacitive behavior of the MOSFET can be examined in each region of operation. Taking PMOS as an example, for a gate-to-substrate voltage V_{GB} that is greater than V_{FB}, the device enters the accumulation region, where further increase in V_{GB} will increase the charge on the gate and cause electrons to accumulate at the oxide-silicon surface. Now assume that V_{GB} is decreased to a value below V_{FB}, shown in Fig. 2-18.

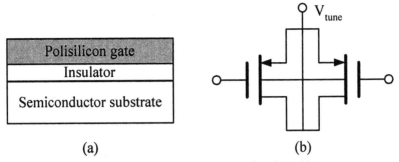

(a)   (b)

Figure 2-16 (a) Two terminal MOS cross-section; (b) MOS varactor

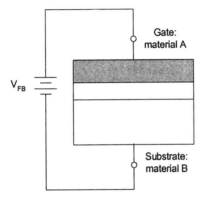

Figure 2-17 Two-terminal MOSFET with external bias V_{FB}

If the difference between V_{GB} and V_{FB} is small, then the potential at the oxide-silicon surface will simply repel the electrons, leaving it depleted of mobile carriers. As V_{GB} continues to decrease, the hole density will eventually exceed the electron density at the surface and an inversion channel with holes forms. If V_{GB} is much smaller than the absolute value of the threshold voltage, then the device enters the strong inversion region. Conventionally, the inversion region is divided into three sub-regions: weak, moderate, and strong. The definition of each sub-region, which is based on the surface potential ϕ_S of the device, is shown in Table 2-1. Note that the moderate and strong inversion region boundaries are not defined "precisely" because they are dependent on the oxide thickness and the substrate doping level. The variable *n* usually takes on values between five and seven [14]. In both strong inversion and the accumulation region, the capacitance of the device is given by:

$$C = \frac{\varepsilon_{ox}}{t_{ox}} \cdot A \qquad\qquad (2\text{-}13)$$

where t_{ox} is the oxide thickness and A is the transistor channel area. When operating in depletion, the weak, and the moderate inversion regions, the capacitance decreases because there are very few mobile carriers at the oxide interface. The capacitance in these three regions can be modeled as the oxide capacitance in series with two parallel capacitances, C_b' and C_i', where C_b' is the depletion region incremental capacitance per unit are, and C_i' is the inversion layer capacitance per unit area [14],[16]. The typical capacitive behavior of the device is shown in Fig. 2-18. The parasitic resistance of the device is approximated with the expression given in [16] and is shown below:

$$R_P \approx \frac{L}{12 k_p W \left(V_{BG} - |V_T| \right)} \qquad\qquad (2\text{-}14)$$

where k_p, W, L, and V_T are gain factor, width, length, and threshold voltage of the device, respectively. The main drawback with such a capacitor is that its behavior is non-monotonic. In order to obtain a monotonic function for the capacitance, the bulk can be disconnected from the drain and source connection and tied to the most positive voltage available, namely V_{dd}. The typical C-V behavior of an inversion-mode MOSFET (I-mode PMOS, or I-MOS) is shown in Fig. 2-19. Even though I-MOS devices provide higher Q and larger capacitance per unit area, they still have issues that are far from desirable. Research has shown that when I-MOS are used for designing voltage-controlled oscillators, simulation and measured center frequencies differ by as much as 20% in the absence of process, voltage and temperature (PVT) variations [16]. A second subtle problem stems from the way that the MOS device is modeled in *SPICE*. A schematic of a MOS transistor model in *SPICE* is shown in Fig. 2-20. The resistors R_S and R_D model the resistances between the source and drain contacts to the metal wires, respectively. Since the MOS transistor is a field-effect device, with the drain and source terminals shorted, the horizontal electric field, and thus the current are both zero. This overly simplistic result prompts the simulator to neglect the channel resistance and optimistically models the MOSFET as an ideal capacitor of value $C_{MOS} = C_{GD} + C_{GS}$. Even though the channel resistance can be inserted into the model, the modulation of this resistance, which is caused by transitions between the sub-regions within the inversion region is still not taken into consideration [16].

Table 2-1 Sub-regions of inversion

Sub-regions	Boundaries
Weak	$\phi_F \leq \phi_S < 2\phi_F$
Moderate	$\sim 2\phi_F \leq \phi_S < 2\phi_F + nV_T$
Strong	$\sim\phi_S \geq 2\phi_F + nV_T$

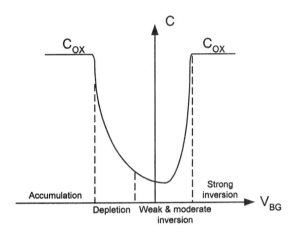

Figure 2-18 Typical tuning characteristic of a two-terminal PMOS

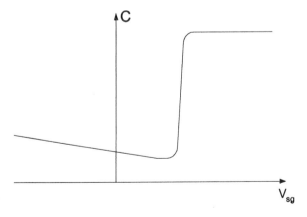

Figure 2-19 Typical tuning characteristic for an I-mode PMOS

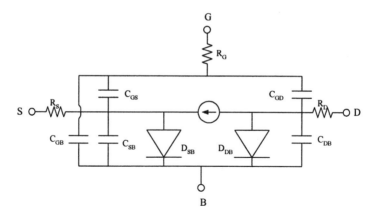

Figure 2-20 MOS transistor *SPICE* model

2.3   Accumulation-mode MOSFET

Rather than operating in the inversion r egion, a better a lternative is to operate the device in the accumulation and depletion regions only [18]. The cross-sectional v iew o f a n a ccumulated-mode M OS (A-MOS) i s shown i n Fig. 2-21. In order to prevent the formation of the inversion regions, the p+ source and drain diffusions are removed from the device and in their places, n+ substrate contact regions are implanted. This not only eliminates the pn-junction capacitance that reduces the tuning range, at the same time this reduces the parasitic n-well resistance.

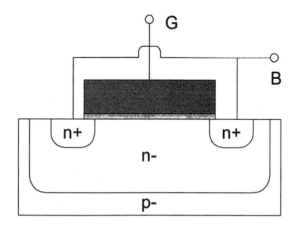

Figure 2-21 Cross-sectional view of A-MOS

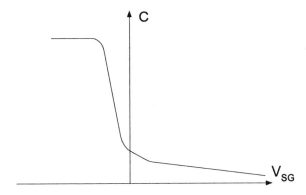

Figure 2-22 Typical tuning characteristic for an A-mode PMOS

The operation of A-MOS is similar to that of the standard MOSFET. The typical C-V curve for an A-MOS is shown in Fig. 2-22 [17]. Even though research has shown that A-MOS provides a wide tuning ability and higher Q [17] – [18], it still has drawbacks. F irst, since it is not really a MOSFET, many processes may not provide a model for it. Second, most layout editors have rules against placing an NMOS in an N-well, so it has a high probability of violating the DRC rules. Instead of placing an NMOS in an N-well, a PMOS can be placed in the P-type bulk to form an A-MOS as well. But usually the performance of such a device is poor since it has relatively low surface mobility.

3. MOS TRANSISTORS

3.1 MOS Transistor High frequency model

Traditionally, CMOS technology is not a good choice for implementing RF circuits because aside from meeting speed, power, and yield requirements that are crucial in digital integrated circuits (IC) design, RF circuits must contend with noise, linearity, and gain as well as power efficiency i ssues [1]. R ecently h owever, w ith the increase o f f_t a nd f_{max} o f CMOS technology as well as the drive towards system-on-chip integration, there has been heavy interest in implementing RF circuits in digital CMOS technology. The advantage of this is quite obvious: if CMOS RF circuits are able to provide similar performance compared to their traditional GaAs and SiGe counterparts, then they may experience their own version of the

Moore's law growth curve just as the microprocessor industry with the integration ease and low fabrication cost of the technology.

In order take full advantage of the performance improvement from scaling the technology, accurate device models are absolutely necessary for circuit simulation [19]. Shown in Fig. 2-23 is a small-signal model of a MOSFET [20]. Note that this is slightly different from the *SPICE* model shown in Fig. 2-20 and from the one used in traditional analog design. Three extra elements are included: channel charge resistance r_i, transconductance delay τ, and gate resistance R_{gi} and R_{ge}.

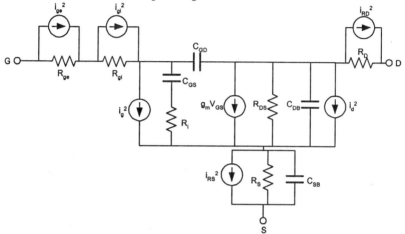

Figure 2-23 Small-signal model of a MOSFET including noise sources

The channel charging resistance models the phenomenon that the channel charge cannot instantaneously respond to changes in the gate-source voltage and is theoretically proven to be $1/5g_m$ [20]. The factor of five comes from the distributed nature of the channel resistance between the source and drain.

The gate resistance is split up into an intrinsic part R_{gi}, and an extrinsic part R_{ge}. The intrinsic resistance is given by the following equation:

$$R_{gi} = \frac{R_g/sq}{3} \cdot \frac{W}{L} \tag{2-15}$$

where R_g/sq is the sheet resistance of the gate material, W and L are width and length of the device, respectively. The factor of three accounts for the distributed nature of the intrinsic gate region [21]. The extrinsic portion of the gate resistance is mainly the contact resistance to the gate material. Most MOSFET models in standard simulator do not take the gate resistance into consideration. This would result in errors when matching to off-chip impedances. Two other crucial issues are that the gate resistance impacts the minimum noise figure as well as the power gain of the device.

Similar to the channel charge resistance, the transconductance delay τ accounts for the fact that the transconductance does not respond to the changes in the gate voltage instantly. Physically, it is the time for the charge to redistribute after an excitation of the gate voltage. Intuitively, is related to the gate-source capacitance of the device and is approximated as [20]:

$$\tau = \frac{C_{gs}}{5g_m} \qquad (2\text{-}16)$$

Before discussing the noise characteristics, a clarification needs to be made between the unity current gain frequency f_t and the maximum oscillation frequency f_{max}. The unity current gain frequency f_t, which is a measure of speed for the technology, is defined as the frequency when the current gain of the device equals unity; i.e., $|y_{21}/y_{11}|=1$. The value of f_t is given in the following expression [20]:

$$f_t = \frac{g_m}{2\pi \cdot \sqrt{\left(C_{GS} + C_{GD}\right)^2 - \left(g_m r_i C_{GS}\right)^2}} \approx \frac{g_m}{2\pi \cdot \left(C_{GS} + C_{GD}\right)} \qquad (2\text{-}17)$$

A crucial observation from Eq. (2-17) is that the current gain is independent o f the gate r esistance. The m aximum oscillation frequency i s defined as the frequency when the power gain delivered by the device equals to unity under matched impedance conditions. I t has a n expression of t he following form:

$$f_{max} \approx \sqrt{\frac{f_t}{8\pi \cdot R_{gi} C_{GD}}} \qquad (2\text{-}18)$$

The key observation here is that since C_{GD} is proportional to the width W of the device, f_{max} is inversely proportional to W if Eq. (2-15) is substituted into Eq. (2-18). The issue that RF circuit designers least like to deal with is probably noise in MOSFETs. This is because noise at one frequency can be modulated by certain circuit operations and down-converted or up-converted to other frequencies. Noise in oscillators and mixers are such examples.

3.2   Noise model of MOS transistor

Three intrinsic thermal noise sources, drain channel noise, gate resistance noise, and induced gate noise exist at RF frequencies while flicker and shot noise exist at frequencies near DC. Drain channel noise is due to the thermal noise generated by carriers in the channel region and is typically represented by:

$$\overline{i_d^2} = 4kT\gamma \cdot g_{do}\Delta f \qquad\qquad (2\text{-}19)$$

where γ is 2/3 for long channel devices and g_{do} is the zero-drain voltage conductance. The parameter γ is dependent on the drain-source voltage. As the drains-source voltage increase, the horizontal electric field increases as well. This causes more carriers to reach velocity saturation, which in turn increases the diffusion noise near the drain [20]. The induced gate noise is originally generated in the channel and then coupled onto the gate as a gate current. At low frequencies, this noise can be neglected. But high frequencies, it can dominate and is expressed as:

$$\overline{i_g^2} = 4kT\delta \cdot g_g\Delta f \qquad\qquad (2\text{-}20)$$

where is δ equals 2γ and g_g is:

$$g_g = \frac{(\omega \cdot C_{GS})^2}{5g_{do}} \qquad\qquad (2\text{-}21)$$

Since the drain channel noise and the induced gate noise arise from the same source, they are correlated. The correlation factor is defined as:

$$\frac{\overline{i_g * i_d}}{\sqrt{\overline{i_g^2} \cdot \overline{i_d^2}}} = jc \qquad\qquad (2\text{-}22)$$

where c = 0.395 for long-channel devices. At low frequencies, the correlation factor c is closer 0.4. As the frequency increases, c is closer to 0.3 [20]. The gate thermal noise, which is generated by the gate resistance, can be expressed as:

$$\overline{v_g^2} = 4kT \frac{R_g/sq}{3} \cdot \frac{W}{L} \Delta f \qquad (2\text{-}23)$$

Shot noise arises in devices due to the result of the discontinuous characteristic of electronic charges that constitute a dc current flowing through reverse-biased pn-junction. The shot noise is very important in BJT case due to DC base current. However, at present, shot noise in MOS transistors can be ignored because the DC gate current is very small. The shot noise is [22]:

$$\left| i_q^{\ 2} \right| = 2qI_{DC}\Delta f \qquad (2\text{-}24)$$

where q is the electronic charge 1.6×10^{-19}C, and I_{DC} is DC bias current.

The last major noise source is the flicker noise, or 1/f noise. It can be found in all active devices and is mainly caused by traps on the silicon-oxide surface. These traps capture and release carriers randomly and give rise to a low-frequency noise signal. Flicker noise is usually modeled as a noise voltage source in series with the gate of the device and is given as:

$$\overline{v_{1/f}^2} = \frac{K_f}{WLC_{OX}} \Delta f \qquad (2\text{-}25)$$

where K_f is a constant for a particular device, typically 3×10^{-12} V²-pF, and C_{ox} is the oxide capacitance per unit area [22]. Even though the flicker noise corner of a MOS transistor is in the kilohertz to megahertz range, flicker noise plays a significant role in the design of oscillators and mixers. These issues are addressed in later chapters.

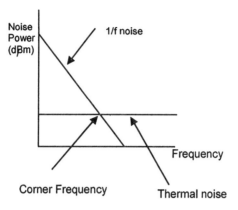

Figure 2-24 Thermal noise and 1/f noise versus frequency

REFERENCES

[1] R. L. Bunch, "A fully monolithic 2.5GHz LC voltage controlled oscillator in 0.35um CMOS technology," *M.S. Thesis*, Virginia Polytechnic Institute and State University, 2001.

[2] S. S. Mohan, "Modeling, design and optimization of on-chip inductors and transformers," *Ph.D Dissertation*, Stanford University, 1999.

[3] A. Niknejad and R. G. Meyer, *Design, simulation and application of inductors and transformers for SI RF ICs*, Kluwer Academic Publishers, Oct. 2000.

[4] H.M. Greenhouse, "Design of planar rectangular microelectronic inductors," *IEEE Trans. on Parts, Hybrids, and Packaging*, pp. 101-109, June 1974.

[5] C. P. Yue and S. S. Wong, "Physical modeling of spiral inductors in silicon," *IEEE Trans. On Electron Devices*, Vol. 47, No. 3, pp. 560 – 568, Mar. 2000.

[6] T. H. Lee, *The Design of CMOS Radio-Frequency Integrated Circuits*, Cambridge University Press, 1998.

[7] J. J. Zhou and D. J. Allstot, "Monolithic transformers and their application in a differential CMOS RF low-noise amplifier," *IEEE J. Solid-State Circuits*, Vol. 33, No. 2, pp. 2020 – 2027, Dec. 1998.

[8] A. Zolfaghari, A. Chan and B. Razavi, "Stacked inductors and transformers in CMOS technology," *IEEE J. Solid-State Circuits*, Vol. 36, No. 4, pp. 620 – 628, Apr. 2001.

[9] J. Long, "Monolithic transformers for silicon RF IC design," *IEEE J. Solid-State Circuits*, Vol. 35, No. 9, pp. 1368 – 1382, Sept. 2002.

[10] C.C. Tang, C.-H. Wu, and S.-I. Liu, "Mminiature 3-D inductors in Standard CMOS Process," *IEEE J. Solid-State Circuits*, Vol. 37, pp. 472 – 480, April 2002.

[11] S. Kodali and D. J. Allstot, "A symmetric miniature 3D inductor," *to be published at ISCAS 2003*.

[12] J. Craninckx and M.S.J. Steyaert, "A 1.8-GHz CMOS low-phase-noise voltage-con-trolled oscillator with prescaler," *IEEE J. Solid-State Circuits*, vol. 30, pp. 1474-1482, Dec. 1995.

[13] Y.-G. Lee, S.-K. Yun, and H.-Y. Lee, "Novel high-Q bond wire inductor for MMIC," *IEEE International Electron Devices Meeting*, pp. 19.7.1-19.7.4, 1998.

[14] Y. Tsividis, *Operation and modeling of the MOS transistor,* McGraw Hill, 1987.

[15] K.E. Mortenson, *Variable capacitance diodes*, Artech House, Inc. 1974.

[16] P. Andreani, and S. Mattisson, "On the use of MOS varactors in RF VCOs," *IEEE J. Solid-State Circuits*, Vol. 35, pp. 905 – 910, June 2002.

[17] P. Andreani and S. Mattisson "A 1.8-GHz CMOS VCO tuned by accumulation mode MOS varactors," *In Proc ISCAS*, pp. 315 – 318, 1998.

[18] T. Soorapanth, C. Yue, D shaeffer, T. H. Lee, S. Wong, "Analysis and optimization of accumulation-mode varactor for RF ICs", *Digest of Technical Papers, Symposium on VLSI Circuits*, pp. 32 - 33, 1998.

[19] Kraus, R and Knoblinger, G, "Modeling the gate-related high-frequency and noise characteristics of deep-submicron MOSFETS," *In Proc. Custom IC Conference*, pp. 209 – 212, 2002.

[20] T. Manku, "Microwave CMOS – Device Physics and Design," *IEEE J. Solid-State Circuits*, Vol. 34, pp. 277 – 285, Mar. 1999.

[21] J. M. Golio, *Microwave MESFET's and HEMT's*. Boston, MA: Artech House, 1991.

[22] P. Gray, P. Hurst, S. Lewis, and R. Meyer, *Analysis and design of analog integrated circuits*, John Wiley & Sons, Inc., 2001.

Chapter 3

PARASITIC-AWARE OPTIMIZATION

In the previous chapter, compact models of passive and active components including key parasitic elements are discussed. The main principle of the parasitic-aware optimization methodology described in this book is that device and package parasitics should be considered at the beginning when the RF circuits are designed as a natural part of the design process. Using the traditional design approach where RF ICs are designed assuming ideal components, it is usually the case that much of the RF circuit performance is when the parasitics are considered due to their degrading effects at high frequencies. The only way to overcome these effects is to consider parasitics in the original design of the circuit. Of course, when the parasitic effects are considered, the complete circuit usually becomes much too complicated for hand analysis so finding the optimum solution is nearly impossible. This motivates the essential need for parasitic-aware design and optimization.

The parasitic-aware optimization methodology comprises three major parts: core optimization tool, CAD simulation tool, and a parasitic-aware model generator. Figure 3-1 shows the overview of parasitic-aware synthesis methodology. The core optimization tool first modifies the *netlist* by varying the design parameters to be optimized in accordance with the chosen optimization algorithm. Then, the *nelist* function invokes the parasitic-aware model generator for the specific *netlist*. The parasitic-aware *netlist* is then simulated by any of several possible circuit simulators specified by the user, such as *HSPICE*, *SPECTRE*, etc. However, not all simulators can be used with the parasitic-aware synthesis paradigm because some simplified simulators do not provide sufficient accuracy to reliably produce an optimum solution. As a consequence of the high-accuracy requirements, the

most time consuming part of the parasitic-aware optimization process is the
circuit simulation within the optimization loop.

The m ost i mportant component of the p arasitic-aware synthesis tool i s
the core optimizer because it controls the overall optimization process much
like the brain controls the human body. It is the details of the core
optimization t ool t hat determine w hether the s olution i s o ptimum p oint o r
merely an undesirable local minimum. Several attractive choices are
available f or p arasitic-aware o ptimization a lgorithms: t he e asiest and most
frequently implemented is the gradient decent search algorithm, while
emerging and more sophisticated methods include simulated annealing,
genetic, and particle swarm algorithms.

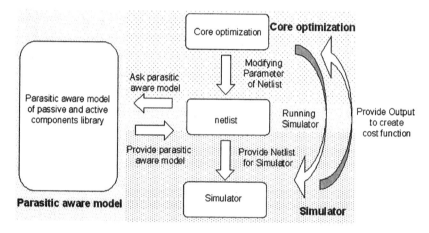

Figure 3-1 Overview of Parasitic-aware optimization

1.   GRADIENT DECENT OPTIMIZATION

Gradient decent search is an algorithm for finding solutions at local
minima that are known to be near the starting point of the simulation. The
gradient optimization algorithm only checks the slope of the cost function: if
the slope is greater than zero, it stops searching for the global optimum
solution and simply reports a local minimum as an optimum value. Because
of this simplicity, it is known to be very fast. However, it has the serious
drawback of easily becoming stuck at undesirable local minima. Figure 3-2
illustrates how this happens for the gradient decent optimization for a simple
one-dimensional cost function. If the starting point of the optimization is in
region I or III, the gradient decent optimizer reports local minimum A or C,

respectively, as the optimum solution, rather than the true global minimum solution at point B. Therefore, most of the potential solutions found by the gradient decent search algorithm are local minima unless the specified starting point falls luckily in region II. Unfortunately, the odds of specifying such an initial condition approach zero for much more complex real-world problems. Figure 3-3 shows a block diagram of the gradient decent optimization computational flow. The maximum number of iterations is set in the first step, and the slope of the cost function is then calculated. If the slope is greater than zero, the solution is rejected, but if the slope is less than zero, it is accepted. This algorithm continues to iterate until all specifications are satisfied or the total number of iterations exceeds the specified maximum number.

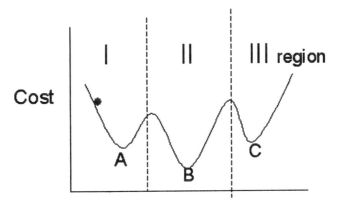

Figure 3-2 Example of gradient decent optimization

2. SIMULATED ANNEALING

The simulated annealing algorithm was introduced to overcome the limitations of the gradient decent optimization. As may be obvious from its name, the concept of simulated annealing optimization follows from the physical process of cooling of metal during manufacture for maximum strength and minimum defects. For example, a metal product such as a knife is heated to a high temperature when it is shaped, and then cooled down slowly in accordance with an empirically determined cooling schedule. This cooling process is termed annealing. After annealing, the metal atoms become stable, and the bonds between them become stronger; in other words, the overall material assumes a minimum energy condition after annealing. The simulated annealing algorithm is designed to fundamentally imitate this process wherein a specified cost function replaces the crystal energy condition of the metal [1].

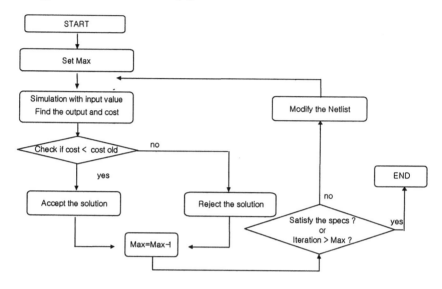

Figure 3-3 Block diagram of gradient decent optimization

Unlike gradient decent algorithms, simulated annealing provides a statistical hill climbing capability, and thus avoids, with high probability, being trapped in undesired local minima. However, the hill climbing process needs to be controlled to save simulation time. In the simulated annealing technique, two control parameters on the hill climbing process are the slope of the cost function and the temp coefficient:

$$rand\ (0,1) < \exp(\ \frac{\text{oldcost} - \text{newcost}}{temp}\ ) \tag{3-1}$$

The left side of Eq. (3-1) is a random number between 0 and 1, and because of the preceding conditional operation shown in Fig.3-4, *oldcost-newcost* (*Δcost*) is always less than 0. Therefore, the right side of Eq. (3-1) is also always less than 1. The condition of *temp* being high with the right side of Eq. (3-1) close to 1 indicates that the probability of hill climbing is high. A second hill-climbing condition is *Δcost*. If that difference is large, the right side of Eq. (3-1) approaches 0 and the hill climbing is much less likely to occur. Table 3-1 summarizes the relationship between the probability of hill climbing and parameters *temp* and *Δcost*.

Table 3-1 Relationship between hill climb probability and parameters temp and Δcost

Probability of Hill Climb	Temp	D(Cost)
High	High	Low
Low	Low	High

In the CAD implementations, *temp* is controlled by α. Initially, temp is set to a high value that decreases according to the coefficient α. Figure 3-4 shows the flow chart of the simulated annealing optimization. Simulated annealing has five different coefficients to set; alpha, beta, *temp*, m and Max. The parameter *temp* is the temperature c oefficient while a lpha is u sed for controlling *temp*. The parameter *m* is the number of iterations with the same *temp* value, and the parameter beta is used for controlling the m coefficient. Max is a maximum number of iterations.

In gradient decent optimization (Fig. 3-2), only the slope of the cost function decides whether solution is accepted or not. However, simulated annealing has a loop for the temperature-cooling algorithm; thus, a second chance is provided to accept the solution according to Eq. (3-1) [2]-[3].

A parasitic-aware RF CAD design and optimization tool should exhibit two key features. First, it should find acceptable solutions efficiently in terms of the required number of computer cycles. Second, it should possess an ability to escape local minima in the design space in order to maximize the probability of finding the global minimum. While the general simulated annealing algorithm is notorious for a hill climbing capability that usually satisfies the second objective, it has several significant shortcomings for parasitic-aware RF circuit synthesis. A critical parameter in determining the cooling schedule in conventional simulated annealing is the *temp* coefficient. Not only is it difficult and time consuming to determine an optimum value for this key empirical parameter, a large number of iterations is required to find a suitable set of acceptable solutions once the value of *temp* is set.

A new simulated annealing heuristic has been proposed with adaptive determination o f *temp* a nd a t unneling p rocess t o s peed u p the g lobal a nd local searches, respectively. Hence, the computer time required for parasitic-aware RF circuit synthesis is reduced and the probability of finding the global minimum is increased.

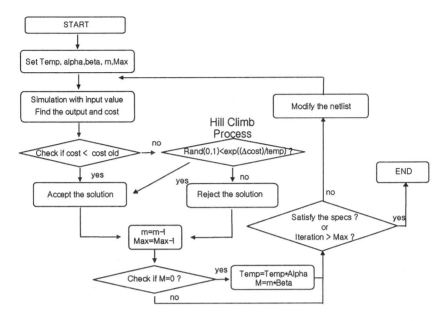

Figure 3-4 Flow chart of parasitic-aware simulated annealing optimization methodology

3.   SIMULATED ANNEALING WITH TUNNELING PROCESS

The new simulated annealing parasitic-aware synthesis approach has three major component parts: the tunneling process, the local optimization technique, and the adaptive *temp* coefficient method.

3.1   Tunneling process

A well-known advantage of simulated annealing is its hill climbing ability, which is determined by the *temp* coefficient in the cooling schedule and the slope of the hill. Showing a simple cost function that depends on a single design variable, X, Fig. 3-5(a) depicts hill climbing in the conventional simulated annealing. There is a finite non-zero probability of escaping local minima in search of the global minimum. However, if *temp* is fixed and the hill is steep, an unacceptably large number of iterations may be required to climb from point A to point B, even though there are no better solutions between those two points as indicated by the higher values of the cost function.

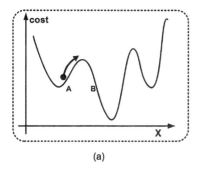 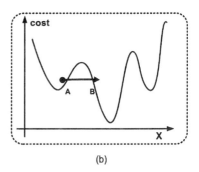

               (a)   (b)

Figure 3-5   (a) Hill climbing in simulated annealing and (b) Tunneling process in adaptive
simulated annealing for a cost function versus design variable, X

In the case of a genetic optimization algorithm, which is described in a
later section, this drawback is often overcome using TABU search methods.
With TABU, unacceptable points are recorded and not simulated again so
the total number of iterations to find an acceptable set of solutions is
reduced. However, the TABU method is not reliably applied with the
simulated annealing algorithm because it effectively inhibits the hill
climbing process. That is, the solution search may become stuck at a local
minimum because TABU disallows hill climbing over known unacceptable
solutions in search of the optimum.

To achieve the advantages of TABU while avoiding the catastrophic
disadvantage described above, a tunneling technique is proposed, as
illustrated conceptually in Fig. 3-7. Rather than wasting time hill climbing
through a region of known unacceptable solutions, or getting stuck, the
optimizer senses the beginning of a climb and enables tunneling from point
A to point B. Whereas TABU considers unacceptable solutions only once,
the tunneling method never considers many unacceptable solutions. Hence,
the required number of iterations needed in parasitic-aware synthesis is
greatly reduced.

Figures 3-6 and 3-7 compare conventional simulated annealing and
adaptive simulated annealing with tunneling. The curves represent values of
a cost function versus values of a single design variable, X. Note that the
cost function has several local minima, and the 'o' and 'x' points portray
accepted and unaccepted solutions, respectively. As illustrated in the
example of Fig. 3-6, conventional simulated annealing wastes iterations
finding unacceptable solutions between points A and B, far from the global
minimum solution. In general, only a small percentage of the solutions are
the desired low cost ones.

In contrast, when the tunneling process is applied as shown in Fig. 3-7,
there are no acceptable and few unacceptable solutions between points A

and B. That is, the searchable design space is reduced because many unacceptable points between A and B are not simulated. Hence, the new tunneling technique focuses on more desirable lower cost acceptable solutions. Thus, there is a higher probability of finding the global minimum quickly without becoming trapped in a local minimum. The initial version of the tunneling algorithm uses coefficients called Tunneling threshold and Tunneling radius. Tunneling threshold determines the conditions when tunneling is invoked, while Tunneling radius specifies the maximum distance a simulation point tunnels in the design space. Tunneling radius is a critical coefficient in the tunneling process. If it is too small, the probability of tunneling through a hill is low; if it is too large, the solution might tunnel so far past the optimum solution that it might never be found. A simple method of determining Tunneling radius is described below.

Figure 3-8 shows a flow chart of the tunneling process. As indicated, when tunneling occurs, a fraction of Tunneling radius is added to the previous parameter values to determine the new design variable values.

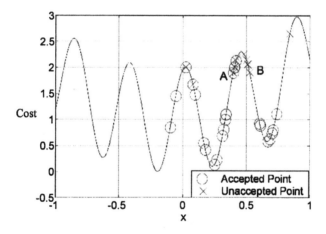

Figure 3-6 Conventional simulated annealing result for a cost function versus design variable, X

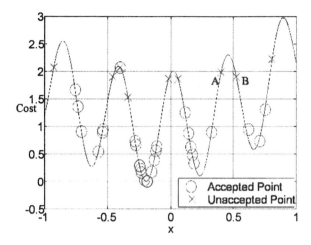

Figure 3-7 Adaptive simulated annealing result for a cost function versus design variable, X

3.2   Local Optimization Algorithm

One major reason that a conventional simulated annealing optimizer requires a large number of iterations to find an acceptable solution is that it chooses its next simulation point randomly. For example, if the optimization problem involves 20 design variables, and each is randomly decreased or increased, then there are $2^{20} = 10^6$ possible directions for the next simulation point! The low probability of choosing a good direction leads to a large number of iterations to find an acceptable solution. On the other hand, if the new direction is set using only information from the previous simulation point, it is easy to become stuck in a local minimum.

As previously mentioned before, the key equation of conventional simulated annealing is

$$rand \; (0,1) < \exp(\frac{oldcost \;\; - \; newcost}{temp}) \tag{3-1}$$

If the coefficient *temp* is large, the probability of hill climbing is high, but if Temp is small, the probability of hill climbing is low.

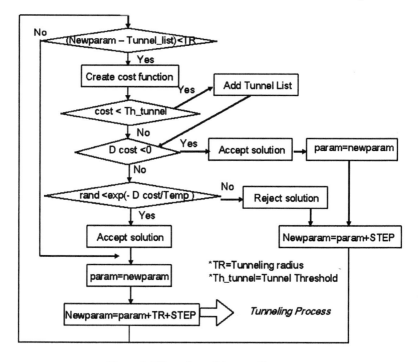

Figure 3-8 Flow chart of the tunneling process

Figure 3-9 depicts the new determination of the direction to the next simulation p oint b ased o n information from both t he p resent a nd p revious points. Vector 'y' is defined between the previous and present points, vector 'x' is chosen randomly, and the new direction is a weighted sum of the two. If *temp* is high, the new direction is predominantly random, but if *temp* low, the new direction is almost identical to 'y'. The next simulation point Z_{NEW} is

$$Z_{NEW} = Z + (F(Temp) \cdot \vec{X} + (1 - F(Temp)) \cdot \vec{Y}) \\ \times \text{Rand}(0.1) \cdot \text{Stepsize} \quad (3\text{-}2)$$

In Eq. (3-2), F (*temp*) is a function that is proportional to the Temp coefficient.

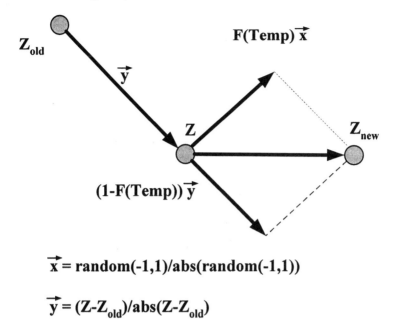

$$\vec{x} = random(-1,1)/abs(random(-1,1))$$

$$\vec{y} = (Z-Z_{old})/abs(Z-Z_{old})$$

Figure 3-9 Weighted vector summations of previous and random directions for the direction to the next simulation point

3.3   Adaptive Temp Coefficient Determination

It has been shown that the value of the *temp* parameter is critical to minimizing the total number of iterations required to find an acceptable solution. If *temp* is too large, iterations are wasted in finding the optimum solution, whereas if *temp* too small, the optimizer may become stuck in a local minimum. Experiences show that the optimum *temp* value is different for different RF circuit topologies. The reason *temp* depends on the circuit is that the cost function slope is different with respect to design space. For example, if the cost function slope of the design space is high, then it requires high temp coefficient to climb the hill of the cost function. On the other hand, if the slope is low, then it requires low temp coefficient to climb the hill of the cost function. To decide the *temp* coefficient adaptively, the slope of the cost function needs to be estimated. The relationship between cost function slope and temp coefficient can be found from Eq. (3-1):

Based on Eq. (3-1), Eq. (3-3) can be derived by setting the random number to $R_{accepted}$ and moving temp coefficient to left side.

$$temp = AVG(\frac{oldcost - newcost}{ln(R_{accepted})}) \qquad\qquad (3\text{-}3)$$

In the proposed adaptive simulated annealing program, by averaging the right side of equation, the slope of the cost function can be estimated and *temp* coefficient can be decided adaptively.

In Eq. (3-3), a new parameter called $R_{accepted}$ that is the initial probability of hill climbing is introduced. Initially, $R_{accepted}$ is set to 70%-90% based on optimization speed. $R_{accepted}$ is a much easier parameter to specify than *Temp* because our experience has shown that it does not depend on the RF circuit topology.

3.4   Comparison between SA and ASAT

A test cost function is chosen as

$$Cost = \cos(14.5 * x - 0.3) + (x + 0.2) * x + 1 \qquad\qquad (3\text{-}4)$$

Figures 3-7 and 3-8 display the test cost function versus values of the single design variable, X. It exhibits several local minima and a global minimum. Figures 3-10 to 3-12 show comparative results between conventional simulated annealing and the new adaptive simulated annealing. The total number of iterations is defined as

$$Total\_iteration = Maxloop * m \qquad\qquad (3\text{-}5)$$

The maximum number of external loops, Maxloop, affects inversely the value of the Temp coefficient. That is, if Maxloop is increased, Temp is decreased. m is the maximum number of iterations with the same value of Temp within a given external loop. As shown in Figure 3-10 and 3-11, the adaptive annealing algorithm has a higher probability of finding a global minimum with a limited number of iterations.

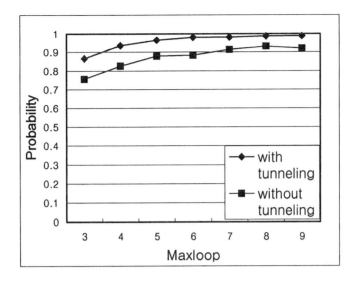

Figure 3-10 Probability of finding the global minimum versus Maxloop for conventional versus adaptive simulated annealing

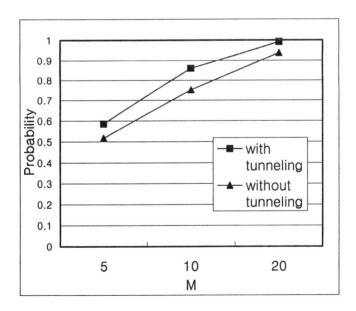

Figure 3-11 Probability of finding the global minimum versus m

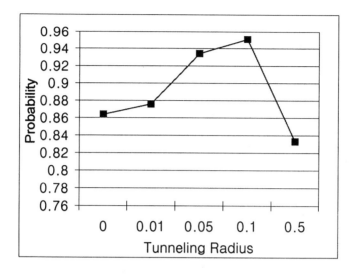

Figure 3-12 Probability of finding the global versus tunneling radius for conventional versus adaptive simulated annealing

Tunneling also introduces a coefficient called *tunneling radius* that determines how far the simulation point jumps when tunneling occurs. This is a critical parameter and is shown in Fig. 3-12; if the radius is too small, tunneling occurs infrequently with results similar to conventional simulated annealing. On the other hand, if the radius is too large, unaccepted solutions might obscure the optimum point so that the global minimum is seldom found. Since process, voltage, and temperature (PVT) variations in an IC range over $\pm 20\%$, we choose 5-10% of the parameter value for the Tunneling radius.

Tunneling process has been applied with a simple cost function and has been shown to be more effective than the conventional hill climbing process in finding the global minimum with a limited number of iterations.

4.   GENETIC ALGORITHM (GA)

The genetic algorithm is one of the most popular optimization algorithms used to find appropriate solutions. The genetic algorithm is inspired by Darwin's theory of evolution; e.g., biological evolution where, over successive generations, individuals who are best suited to survive in an environment live on and reproduce, while other individuals die off. This algorithm has proliferated since John Holland and his colleagues invented it in the 1970's and hence there are many different types of GAs used today. This algorithm has also been applied to analog and RF synthesis [4],[5].

The genetic algorithm starts with a population of generated solutions, *chromosomes*, which can be random or chosen to be close to the desired solution. The *chromosomes* are usually encoded in bit strings, and each bit or group of bits represents a *gene* as shown in Fig. 3-13. After the first generation is created, the *chromosomes* are decoded into their actual representation, analyzed and given a fitness value to characterize how close to the ideal solution they reside.

Based on the calculated fitness values, a s ubset of the chromosomes is selected to mate and reproduce so that a new generation is created. *Selection* can take many different forms. Among many different selection methods, roulette wheel selection is most popularly used. In roulette wheel selection, each chromosome i s given a p roportion o f t he r oulette w heel e qual t o the proportion of its fitness value to the sum of all fitness values of the generation. Then a roulette wheel is spun a number of times and the chromosome whose slot the ball lands in is selected to reproduce. The best chromosome is usually added to the next generation in order not to lose the best solution through the generations.

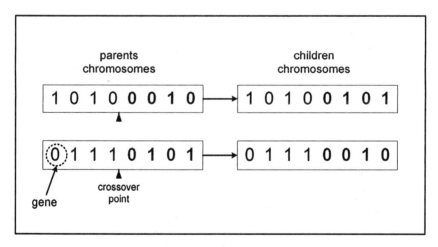

Figure 3-13 Genetic algorithm: *selection* and *crossover*

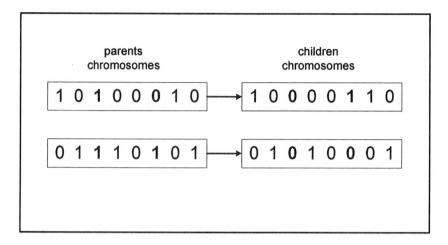

Figure 3-14 Genetic algorithm: *mutation*

Chromosomes reproduce by a process called *crossover*. During this stage, two parents are chosen randomly from the set of selected chromosomes and each chromosome is cut into two at the same location. The halves are interchanged to create two children chromosomes to be included in the next generation. The parent chromosomes may or may not also be included in the next g eneration. C rossover c an a lso b e p erformed w ith m ultiple c uts, t hus allowing genes in the middle of the chromosome to be interchanged.

As a final step, some chromosomes may have their genes mutated to avoid the set of solutions from be stuck in local minima (Fig. 3-14). *Mutations* can include a simple bit inversion in the chromosome or other more complex operations such as complete gene modifications.

These processes of *selection*, *crossover*, and *mutation* are repeated until a chromosome with a fitness value close to the desired value is found or until a predefined maximum number of generations have been generated. The chromosome with the best fitness value at the end of the optimization process is decoded into the optimum solution. The overall genetic algorithm process is summarized in Fig. 3-15.

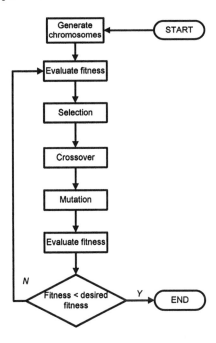

Figure 3-15 Flow chart of genetic algorithm

5. PARTICLE SWARM OPTIMIZATION (PSO)

This evolutionary computational technique was originally inspired from observations of group social behavior, e.g., a flock of birds flying in search for the best location of food [6]. Since this optimization algorithm works concurrently on a population of potential solutions rather than a single one unlike simulated annealing, it is generally k nown to b e robust and fast in finding an optimum solution. It is also different from other population-based optimization algorithms in that it has a unique *swarming* process.

5.1 Particle swarm optimization algorithm theory

A number of social scientists found that *swarming* differs from other group behaviors of animals in that individual members of the group can profit from the discoveries and previous experience of all other members of the group during the search for food [7]. The swarming behaviors are usually found in a flock o f b irds, butterflies, a nd a sc hool o f f ish. P article swarm algorithm is basically simulating this swarming behavior and has been implemented successfully in a number of problems requiring optimization.

Particle swarm algorithm works on a population of potential solution candidates called *particles*. The movements of the group of particles are governed by updating the position and velocity of individual particle in an effort to find the optimum solution. Therefore the key of success of the algorithm is how effectively to move each individual particle through the multi-dimensional design hyperspace.

$$\vec{x}(t+1) = \vec{x}(t) + \vec{v}(t)$$

$$\vec{v}(t+1) = \overbrace{w.\vec{v}(t)}^{\text{Inertia}} + \overbrace{c_1.rand(0,1).(\vec{x}_{Pbest}(t) - \vec{x}(t))}^{\text{Competition}}$$
$$+ \underbrace{c_2.Rand(0,1).(\vec{x}_{Gbest}(t) - \vec{x}(t))}_{\text{Cooperation}}$$

Figure 3-16 Particle swarm equations: position and velocity of each particle

The obvious advantage of the algorithm is its simplicity. It is conceptually easy to understand and very simple to implement since the optimization algorithm can be explained by only two straightforward equations as shown in Fig. 3-16.

Generally, any motion of an object can be expressed by two vector components: present position and present velocity. The position of this object at time t is simply a vector sum of these two. Similarly, PSO uses present position and velocity vector components to determine the next position of each particle. This is described in the first line of the equations. The second line illustrates how to determine the next velocity. The velocity update process consists of three different vectors called *inertia, competition,* and *cooperation*, respectively. *Cooperation* is the vector between the globally best position that one member of the group had found and the current position of the particle with weighted by a uniformly distributed random function. Since each particle should have knowledge of the globally best position (x_{Gbest}) by cooperating and communicating each other, this factor is called *cooperation* factor. *Competition* is the vector between the personally best position (x_{Pbest}) that each individual member had experienced and the current position. It is called *competition* factor since it describes the tendency of each particle to explore the personally experienced best position more and it doesn't need any communication with others unlike the *cooperation* factor. Finally, *inertia* is simply the previous velocity weighted by a constant, w. Since it is the tendency to maintain the previously moving direction, it is called *inertia*. Particle swarm algorithm combines these factors, namely, *inertia*, *competition*, and *cooperation*.

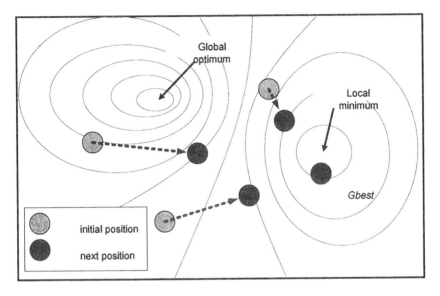

Figure 3-17 Optimization process with only *cooperation* factor

At a glance of this equation, it is unclear why it needs *inertia* and *competition* factors while *cooperation* is undoubtedly indispensable to finding the better location of the optimum solution by its definition. The main idea of adding inertia and competition is avoiding trapping in local minima. Let's examine a simple example in Fig. 3-17, which demonstrates the algorithm with only the *cooperation* factor. The probability of finding the global optimum without being trapped in local minima is strongly dependent on the initial location of each particle because the particles always try to explore only the vicinity of the position known to be the best so far. This is very similar to the gradient optimization as described in the previous section. This problem is solved using the complete particle swarm algorithm as shown in Fig. 3-18. It increases the probability to find the global optimum without becoming trapped in local minima. In addition to increased robustness, it also improves the speed mainly because of the *competition* factor.

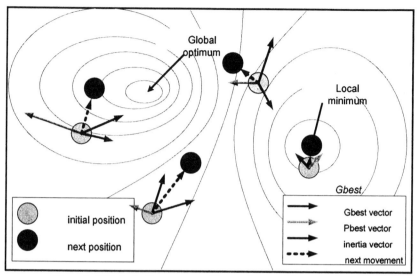

Figure 3-18 Optimization process with complete particle swarm formula

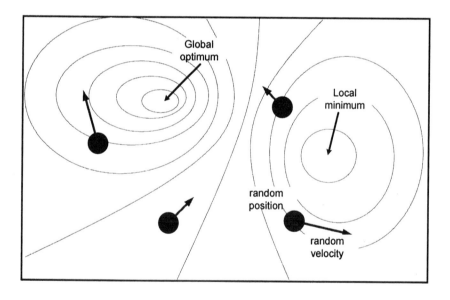

Figure 3-19 System initialization with random position and velocity (number of particles: 4)

The way that these three factors are combined is very critical in determining the efficiency and the robustness of the algorithm. It is reasonable to assign the same weight for each factor in general since the problem is initially assumed to be unknown; in other words, the possibility of getting c lose to the g lobal s olution c ontributed by each factor is s ame. Therefore, the weights for the factors, w, $c_1 rand(0,1)$, and $c_2 Rand(0,1)$ are chosen to be 1 in average and hence c_1 and c_2 are 2. However, the inertia weight, w, is sometimes adjusted to be less than 1 to make the particles converge in time. This will be discussed in more detail in the following sections.

Because of the simplicity of the algorithm, it requires only primitive mathematical operators, and is computationally inexpensive in terms of both memory requirements and speed.

5.2   Optimization procedure

In particle swarm, the system is initialized with a population of random solutions. Each potential solution is also assigned a random velocity, and the potential solutions, *particles*, are then flying through the problem space as shown in Fig. 3-19. Each *particle* also keeps track of the best solution it has experienced, which is called *Pbest*. The whole group of *particles* also keeps track of the best solution among the group called *Gbest*. Therefore the particles converge toward the solution based on *Pbest* experience and *Gbest* experience by changing the velocity of each particle at each time. This process repeats until some condition is satisfied (for example, until a solution with a cost value close t o the desired value or until a predefined maximum number of iterations has been reached). The overall optimization procedure is summarized in Fig. 3-20.

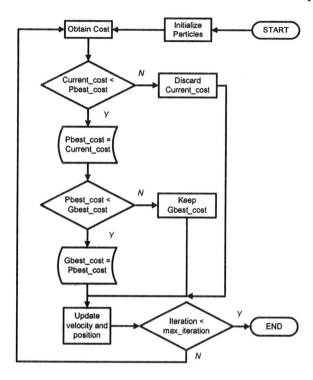

Figure 3-20 Flow chart of particle swarm optimization for CMOS RF ICs

5.3   Optimization parameters

The inertia weight, w, is a very critical parameter to decide the balance between the global and the local search abilities. Since this parameter works as a resistance force, in a sense, to converge on the presently known personal and group best positions, the optimizer will focus more on the global search if w is too large and hence make the optimizer unnecessarily slow. On the other hand, if w is too small, it will have more local search ability and hence the chance to be trapped in local minima will be increased.

Another important optimization parameter, which is not shown in Fig. 3-16, is v_{max}. It is a parameter that limits the maximum velocity that particles can fly with. It decides the dynamic range of particles at each time step.

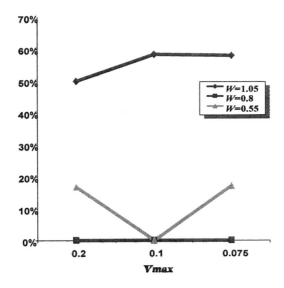

Figure 3-21 Failure rate with different parameter settings

In order to investigate the impact that the parameters of particle swarm have on performance of the parasitic-aware RF design and optimization, an RF power amplifier synthesis is demonstrated as a test vehicle. The detail design and specification of these circuits is explained in Chapter 7. The same circuit has been synthesized 12 times using different parameter settings for statistical comparison, and the failure rate (the number of synthesized circuits that could not find the optimum solution divided by 12) and the average number of iterations needed to find the optimum have been recorded in Fig. 3-21 and Fig. 3-22, respectively.

From this results, $w = 0.8$ and $v_{max} = 0.1$ outperforms others in both failure rate and average number of iterations. Other cases imply imbalance between the global and the local search abilities, and hence result in poor robustness in addition to the slow convergence of the optimization process. This setting has been shown to work very well in other circuits as well [8].

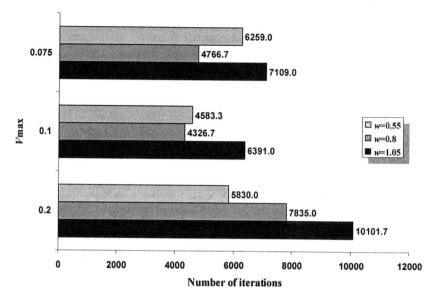

Figure 3-22 Average number of iterations with different parameter settings

6.   POST PVT VARIATION OPTIMIZATION

One of the key unresolved issues in optimization is how to optimize a
design in the presence of process, voltage and temperature (PVT) variations.
One method is to add the PVT variations to the cost functions. To consider
PVT variations requires more than 1 SPICE simulation per iteration cycle
because of the many different temperature, voltage and process parameter
values for each potential solution. So, an unacceptably large number of
iterations are required to find the optimum solution. For example, if 200
simulations are required to get PVT variation information and 10000
iterations are required to find the optimum solutions without considering
PVT variations, the total iteration count is $200 \times 10000 = 2000000$!

An alternative approach, called post PVT optimization, is to run PVT
simulations with only the accepted solutions. The optimization program
needs to maintain a tunnel list capability so that it avoids simulating the
same points over and over again.

Fundamentally, PVT optimization involves running Monte Carlo
simulations with different temperature, voltage and process parameters, and
then redefining the optimum solution once the average and standard

deviations of the cost function are found. This process does not necessarily lead to the same solution as obtained without PVT consideration.

If after PVT simulations, the solution found does not satisfy the specifications, then these solutions are added to the tunnel list to avoid re-simulation; the optimization program is then run again until the solution meets the specifications. Figure 3-23 shows a block diagram of simulation techniques using PVT variations. When the optimum solution does not satisfy the specifications for the first time, the adaptive simulated annealing is run again. However, subsequent runs do not take as long because the tunnel list has excluded the sub-optimum solutions.

The post optimization idea can also be applied to circuits that require a lot of time to simulate, such as VCOs, which need to run time-consuming phase noise simulations. In this particular case, a cost function in the main program can check if the VCO oscillates at the correct frequency. In the post optimization phase, phase noise and tuning range simulations can be added and thus a lot of simulation time is saved.

In Chapter 7, an example of the post PVT optimization is discussed.

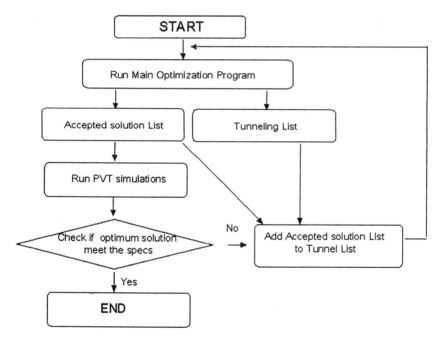

Figure 3-23 Block diagram of post PVT variation optimization

REFERENCES

[1] R.A. Rutenbar, "Simulated annealing algorithms: An overview," *IEEE Circuits and Devices Magazine*, pp. 19-26, Jan. 1989.

[2] R. Gupta and D.J. Allstot, "Parasitic-Aware Design and Optimization of CMOS RF Integrated Circuits," *IEEE RFIC Symposium*, pp. 325-328, June 1998.

[3] R. Gupta and D.J. Allstot, "Parasitic-Aware Design and Optimization of CMOS RF Integrated Circuits, *IEEE International Microwave Symp.*, pp. 1847-1850, June 1998.

[4] W. Kruiskamp and D Leenaerts, " Darwin: Cmos opamp synthesis by means of a genetic algorithm," *IEEE Design Automation Conference*, pp. 433-436, 1995.

[5] P. Vancorenland, G. Van der Plas, M. Steyaert, G. Gielen, and W. Sansen, "A layout-aware synthesis methodology for RF circuits, " *ICCAD*, pp. 358-362, 2001.

[6] J. Kennedy, et al., "Particle swarm optimization," *Proc. IEEE Int. Conf. on Neural Networks*, vol. 4, pp. 1942-1948, 1995.

[7] R. Eberhart, et al., "A new optimizer using particle swarm theory," *IEEE Int. Symp. On Micro Machine and Human Science*, pp. 39-43, 1995.

[8] J. H. Park, K. Y. Choi, and D. J. Allstot, "Parasitic-aware design and optimization of a fully integrated CMOS wideband amplifier," *to be published at ASP-DAC*, 2003.

PART II: OPTIMIZATION OF CMOS RF CIRCUITS

Chapter 4

OPTIMIZATION OF CMOS LOW NOISE AMPLIFIERS

1. LOW NOISE AMPLIFIER

The low noise amplifier is the first block in the receiver chain and is used to amplify the weak RF signal arriving from the external antenna, duplexer switch, and band select filter. Since the signal from the antenna is very weak, the amplifier is required to add small noise yet provide high gain; hence, the name low noise amplifier. To minimize the various noise contributions of the amplifier circuits, the noise source needs to be carefully analyzed and subsequently optimized.

1.1 Noise

Two different types of noise sources exist in low noise amplifiers: thermal noise generated by resistors, including parasitic resistances, and noise generated by the active devices.

1.1.1 Thermal noise

The basic thermal noise equations are (Fig. 4-1) [1]

$$\left| I^2 \right| = 4kTG\Delta f \qquad (4\text{-}1)$$

and

$$\left|V^2\right| = 4kTR\Delta f \tag{4-2}$$

where k is Boltzmann's constant (1.38×10^{-23} J/K), T is the absolute temperature, G is the conductance of the resistor, and Δf is a bandwidth of interest.

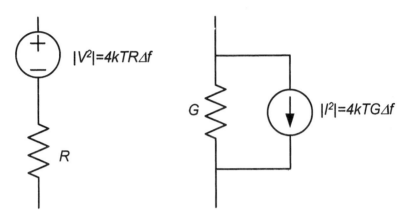

Figure 4-1 Resistor thermal noise model

For example, if the resistance is 50 ohm, temperature is 300K, and the bandwidth is 1MHz, then the mean square open circuit noise voltage is

$$\left|V^2\right| = 4kTR\Delta f$$
$$= 4 \cdot 1.38 \times 10^{-23} (J/K) \times 300(K) \times 50(ohm) \times 1 \times 10^6 (Hz) \tag{4-3}$$
$$= 8.28 \times 10^{-13} (V^2) = -138(dBm)$$

The mean square short circuit noise current is

$$\left|I^2\right| = 4kTG\Delta f$$
$$= 4 \cdot 1.38 \times 10^{-23} (J/K) \times 300(K) \times \frac{1}{50}(Siemen) \times 1 \times 10^6 (Hz) \tag{4-4}$$
$$= 3.31 \times 10^{-16} (A^2) = -138(dBm)$$

Since the thermal noise depends on the bandwidth rather than any absolute frequency, the thermal noise spectrum density is independent of frequency. If the noise is independent of the frequency it is called a white noise because it contains all components of the frequency spectrum. For noise in MOS transistors, the readers may refer to Chapter 2 of this book.

1.1.2   Noise figure

To evaluate the noise performance of a low-noise amplifier, the equation for noise figure is given by

$$NF = 10\log(F) = 10\log(\frac{SNR_{in}}{SNR_{out}}) = 10\log(\frac{\frac{signal}{Noise_{in}}}{\frac{signal}{Noise_{out}}})$$

$$= 10\log(\frac{Noise_{in} + Nosie_{amp}}{Noise_{in}}) = 10\log(1 + \frac{Noise_{amp}}{Noise_{in}})$$

(4-5)

where F is noise factor and SNR_{in} is the signal-to-noise ratio at the input and SNR_{out} is the signa-to-noise ratio at the output. $Noise_{in}$ is the noise from the previous stage, $Noise_{out}$ is the noise at the output which consists of the noise from amplifier ($Noise_{amp}$) plus the noise from Noise$_{in}$. For example, if the noise contributed by the previous stage is identical to the noise contributed by the low-noise amplifier, then the NF is 3 dB.

Why is a low noise amplifier needed in an RF receiver chain? According to the Friis equation [5], when blocks such as an LNA, Mixer, and filter, etc, are cascaded in a receiver (Fig. 4-2), the overall, noise factor F is

$$F = 1 + (F_1 - 1) + \frac{(F_2 - 1)}{A_{p1}} + \frac{(F_3 - 1)}{A_{p1}A_{p2}} + \dots$$

$$F_n = F_1 + \frac{(F_2 - 1)}{A_{p1}} + \dots + \frac{(F_n - 1)}{\prod_{i=1}^{n-1} A_{pi}}$$

(4-6)

As shown in Eq. (4-6), the first stage noise factor is critical in determining the overall noise factor of the system because the noise factors of the second and subsequent stages are divided by the previous stage gains when referred back to the input. Hence, the LNA is required to have high gain and low noise to achieve an overall noise factor that is sufficiently low.

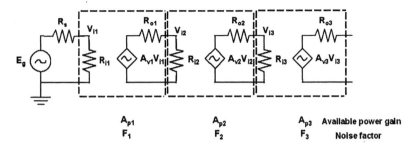

A_{p1} A_{p2} A_{p3} Available power gain
F_1 F_2 F_3 Noise factor

Figure 4-2 Noise model of Cascade system

1.2   Linearity

There are two different limitations in determining the dynamic range of the amplifier (Fig. 4-3). One is noise and the other is linearity. Usually, noise figure sets the limitation on minimum signal strength, and linearity limits the maximum signal strength. Hence, linearity is equally as important as noise figure in the design of a low noise amplifier. In the case of base band amplifiers, total harmonic distortion is usually used to represent their linearity. On the other hand, RF amplifiers often use third-order input intermodulation distortion (IIP3), second-order input intermodulation distortion (IIP2), or the 1 dB gain compression point to represent their linearity.

1.2.1   1dB gain compression point

Since the amplifier obviously cannot be perfect, the output signal after amplification contains the desired as well as undesired signals including fractional harmonic distortion terms. Eq. (4-7) shows a polynomial representation of a weakly nonlinear amplifier.

$$S_o(t) \approx a_1 S_i(t) + a_2 S_i^2(t) + a_3 S_i^3(t) \quad \text{.......}$$

$$\text{If} \quad S_i(t) = S_1 \cos \omega_1 t$$

$$S_o(t) = a_1 S_1 \cos \omega_1 t + a_2 S_1^2 \cos^2 \omega_1 t + a_3 S_1^3 \cos^3 \omega_1 t$$

$$= a_1 S_1 \cos \omega_1 t \qquad\qquad (1)$$

$$+ a_2 S_1^2 (1/2)(\cos 2\omega_1 t + 1) \qquad (2)$$

$$+ a_3 S_1^3 (1/4)(\cos 3\omega_1 t + 3 \cos \omega_1 t) \qquad (3)$$

(4-7)

As seen above, the amplifier transfer function contains several high-order amplification terms that cause the harmonic and other nonlinear output components. Term (1) of Eq. (4-7) shows the fundamental desired signal and term (2) shows the second-order term that consists of a second harmonic distortion signal plus DC offset signal. Likewise, term (3) shows a third order harmonic distortion term plus a signal at the fundamental frequency that therefore represents gain compression or expansion. In general, the even order amplification terms create even harmonic distortion components as well as DC offset bias point shifts, while the odd order terms cause odd harmonic distortion as well as well as gain compression or expansion.

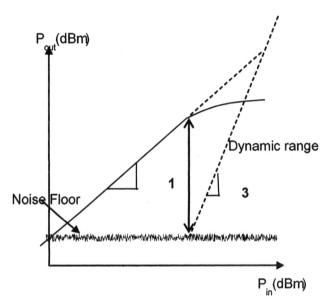

Figure 4-3 Spurious-free dynamic range with noise floor and IIP3

Usually, the coefficient a_3 is negative, so it causes the fundamental signal gain compression which leads to the definition of the related 1 dB gain compression point as shown in Fig. 4-4. Specifically, the 1 dB gain compression point is defined as the input power level in dBm at which the overall gain of amplifier is reduced by 1 dB from its maximum value.

According to the definition, the 1 dB gain compression point is

$$20\log(a_1 S_1 + \frac{3}{4} a_3 S_1^{\;3}) = 20\log(a_1 S_1) - 1dB$$

$$a_1 S_1 + \frac{3}{4} a_3 S_1^{\;3} = \frac{a_1 S_1}{1.122} \tag{4-8}$$

$$0.1087 a_1 = -\frac{3}{4} a_3 S_1^{\;2}$$

If a_3 is negative, then the magnitude of the required input voltage S_1

$$S_1 = V_{in} = \sqrt{0.145 \left|\frac{a_1}{a_3}\right|} \tag{4-9}$$

Pin_{1dB} is the input power in dBm. If the input is perfectly matched to the source impedance, and R_{in} is 50 ohm,

$$Pin_{1dB} = 10\log(\frac{V_{in}^{\;2}}{R_{in}}) + 30 = 10\log(V_{in}^{\;2}) + 30 - 10\log(50)$$

$$= Vin(dB) + 30 - 17 = 10\log(0.145(\frac{a_1}{a_3})) + 13 \tag{4-10}$$

$$= 13 + 10\log(0.145) + 10\log(a_1) - 10\log(a_3)$$

$$= 4.6137 + 10\log(a_1) - 10\log(a_3)$$

1.2.2   Two-tone test (IIP2 and IIP3)

To measure the linearity of an RF block, the two-tone test is commonly used as shown in Fig. 4-5 where one of the tones represents the desired RF signal and the other represents an adjacent channel interfering signal. The two-tone test is used to check for the in-band inter-modulation terms resulting from the application of these signals. Usually, IIP3 and IIP2 are measured using the two-tone test method. The equation representing this test is:

$$S_o(t) \approx a_1 S_i(t) + a_2 S_i^2(t) + a_3 S_i^3(t) \quad \ldots\ldots$$

If $S_i(t) = S_1 \cos\omega_1 t + S_2 \cos\omega_2 t$

$$S_o(t) = a_1(S_1 \cos\omega_1 t + S_2 \cos\omega_2 t) + a_1(S_1 \cos\omega_1 t + S_2 \cos\omega_2 t)^2$$

$$+ a_3(S_1 \cos\omega_1 t + S_2 \cos\omega_2 t)^3 \quad \ldots\ldots$$

$$= a_1 S_1 \cos\omega_1 t + a_1 S_2 \cos\omega_2 t \tag{1}$$

$$+ a_1 S_1 S_2 \cos(\omega_1 + \omega_2)t + a_1 S_1 S_2 \cos(\omega_1 - \omega_2)t \tag{2}$$

$$+ \frac{3a_3 S_1^2 S_2}{4}\cos(2\omega_1 + \omega_2)t + \frac{3a_3 S_1^2 S_2}{4}\cos(2\omega_1 - \omega_2)t \tag{3}$$

$$+ \frac{3a_3 S_1 S_2^2}{4}\cos(2\omega_2 + \omega_1)t + \frac{3a_3 S_1 S_2^2}{4}\cos(2\omega_2 - \omega_1)t \tag{4}$$

$$(4\text{-}11)$$

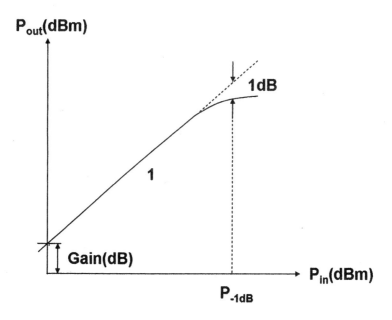

Figure 4-4 1dB compression point

As shown in Eq. (4-11), term (1) is the desired fundamental signal, term (2) represents the possible second-order inter-modulation terms of $\omega_1 \pm \omega_2$, terms (3) and (4) are possible third-order inter-modulation terms at $2\omega_1 \pm \omega_2$ and $\omega_1 \pm 2\omega_2$. The second term in (2) has a frequency values close to DC, so it contributes to the overall DC offset. Since the frequencies of the second terms in (3) and (4) are close to desired signal frequency, these inter-modulation terms cannot be filtered out without adversely affecting the desired signal. However, the first terms in (1), (2), and (3) are easily filtered away. IIP2 as represented by the second term in (2) is a particular problem in homodyne receivers, whereas the IIP3 represented in (3) and (4) are problematic for heterodyne receivers.

The definition of IIP2 is the input power in dBm where the fundamental output power and the second-order intermodulation output power are the same (Fig. 4-5 (b)). According to the definition, the equation for IIP2 in voltage is

$$20\log(a_1 S_1) = 20\log(a_2 S_1^{\,2})$$

$$a_1 S_1 = a_2 S_1^{\,2} \tag{4-12}$$

$$S_1 = V_{in} = \frac{a_1}{a_2}$$

IIP2 is the input power in dBm. If the input is perfectly matched, and R_{in} is 50 ohm then,

$$IIP2 = 10\log(\frac{V_{in}^{\,2}}{R_{in}}) + 30 = 10\log(V_{in}^{\,2}) + 30 - 10\log(50)$$

$$= Vin(dB) + 30 - 17 = 10\log((\frac{a_1}{a_2})) + 13 \tag{4-13}$$

$$= 13 + 10\log(a_1) - 10\log(a_2)$$

The definition of IIP3 is the input power in dBm where the fundamental output power and the third-order intermodulation output power are the same (Fig. 4-5 (c)). According to the definition, the equation of IIP3 is

$$20\log(a_1 V_{in}) = 20\log(\frac{3}{4}a_3 V_{in}{}^3)$$

$$a_1 V_{in} = \frac{3}{4}a_3 V_{in}{}^3 \tag{4-14}$$

$$V_{in}{}^2 = -\frac{4}{3}\frac{a_1}{a_3}$$

If a_3 is negative, then

$$V_{in} = \sqrt{\frac{4}{3}\left|\frac{a_1}{a_3}\right|} \tag{4-15}$$

IIP3 is the input power in dBm. If the input is perfectly matched, and Rin is 50 ohm then,

$$
\begin{aligned}
IIP3 &= 10\log(\frac{V_{in}{}^2}{Rin}) + 30 = 10\log(V_{in}{}^2) + 30 - 10\log(50) \\
&= Vin(dB) + 30 - 17 = 10\log(\frac{4}{3}(\frac{a_1}{a_3})) + 13 \\
&= 13 + 10\log(\frac{4}{3}) + 10\log(a_1) - 10\log(a_3) \\
&= 14.2494 + 10\log(a_1) - 10\log(a_3)
\end{aligned}
\tag{4-16}
$$

The relation between IIP3 and the 1dB compression point is

$$IIP3 - Pin_{1dB} = 14.2494 - 4.6137 = 9.6\ dB \tag{4-17}$$

If two blocks are cascaded, then the overall IIP3$_{TOT}$ is

$$\frac{1}{IIP3_{TOT}} = \frac{1}{IIP3_2} + \frac{A_2}{IIP3_1} \tag{4-18}$$

where A_2 is the available power gain of second stage, $IIP3_1$ is the first stage IIP3, and $IIP3_2$ is the second stage IIP3. As shown in Eq. (4-18), if the gain of A_2 is large enough, the overall IIP3 is dominated by the last stage IIP3.

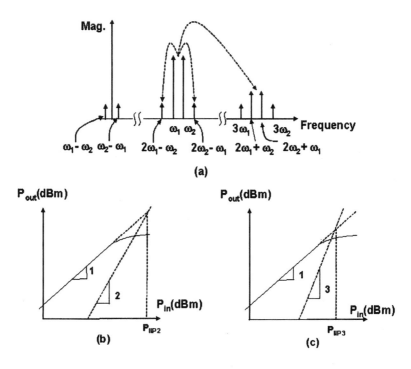

Figure 4-5 (a) Frequency spectrum of two-tone test; (b) IIP2 and (c) IIP3

2.   DESIGN OF LOW NOISE AMPLIFIER

Typical specifications of a low-noise amplifier are:

NF = 3 dB
IIP3 = -10 dB
Gain = 20 dB
Input and Output matching < -10 dB
Stability factor >1
Bias current <4 mA

where the Stern stability factor (K) is defined as

$$K = \frac{1+|\Delta|^2 - |S_{11}|^2 - |S_{22}|^2}{2|S_{21}||S_{22}|}$$

(4-19)

where $|\Delta|$ is

$$|\Delta| = |S_{11}S_{22} - S_{12}S_{21}|$$

(4-20)

In the design of the LNA, the gain, noise figure and linearity specifications should be met with the minimum possible power dissipation. Also, overall stability and impedance matching at the input and output terminals must also be considered. The most popular RF LNA topology for relatively low-frequency applications is the common-source cascode amplifier with inductive source degeneration as shown in Fig. 4-6. The advantage of the cascode amplifier is that it can increase reverse isolation due to the suppression of the miller capacitance between amplifier input and output terminals. In addition, the cascode amplifier is made more stable.

The advantage of inductor source degeneration is that it can achieve a real input impedance component without adding thermal noise. The input impedance of the LNA is

$$Z_{in} = s(L_g + L_s) + \frac{1}{sC_{gs}} + \left(\frac{g_m}{C_{gs}}\right)L_s$$

(4-21)

So, by using real term of Z_{in}, we can minimize the input reflection and maximize gain.

The corresponding noise factor equation is

$$F \approx F_{min} + \frac{Rn}{Gs}(Gs - Gopt)^2$$

(4-22)

where F_{min}, Rn, and $Gopt$ are

$$F_{min} = 1 + 2.4\frac{\gamma}{\alpha}\left[\frac{\omega}{\omega_T}\right]$$

(4-23)

$$R_n = \frac{\gamma}{\alpha} \cdot \frac{1}{gm} \tag{4-24}$$

$$Gopt = \alpha \omega C_{gs} \sqrt{\frac{\delta}{5\gamma}(1-|c|^2)} \tag{4-25}$$

So, depending on the Gs, noise factor can be changed as shown in Fig. 4-7. One the other hand, the transducer power gain of the amplifier is

$$A_{TU} = A_S A_O A_L = \frac{1-|\Gamma_S|^2}{|1-S_{11}\Gamma_S|^2}|S_{21}|\frac{1-|\Gamma_L|^2}{|1-S_{11}\Gamma_L|^2} \tag{4-26}$$

According to the Eq. (4-26), the first term of the gain depends on the input matching condition while the second and third terms depend on S_{21} of the amplifier and its output matching condition S_{22}.

So, to design an input matching network, two factors need to be considered. One is the maximum power transfer condition and the other is noise matching to minimize the overall noise figure. Figure 4-7 is an example that shows the trade off between noise a nd gain due to the input matching. For example, for maximum gain which is about 20 dB, the noise figure will be greater than 4.5 dB. On the other hand, if the input is matched for the minimum noise figure, which is 1 dB, the gain drops to only 2 dB. So, based on the specifications, the gain and noise figure need to be traded off in the design and optimization process.

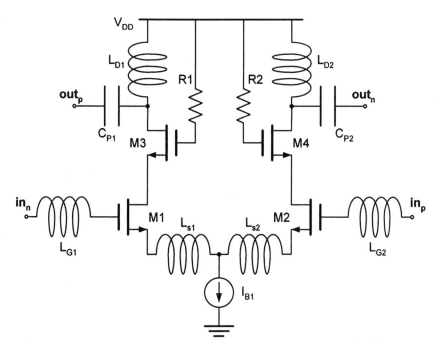

Figure 4-6 Differential cascode common-source LNA with source degeneration inductors

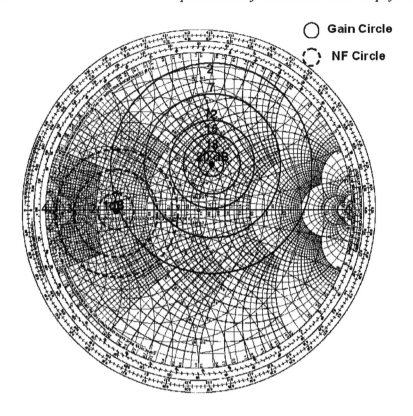

Figure 4-7 Trade offs between gain and NF in Smith chart

3.   OPTIMIZATION OF LOW NOISE AMPLIFIERS

3.1   Calculating gate induced noise in SPICE

To design the RF low noise amplifier, gate induced noise is often very critical. However, conventional CMOS transistor models do not always model gate induced noise as well as the correlation between gate and drain noise. To consider gate induced noise, first the gate noise which is blue noise needs to be changed to white noise to use a regular resistor (Fig. 4-8). According to the Eq. (4-27), the gate induced noise is [4]

$$g_g = \frac{\omega^2 C_{gs}^2}{5 g_m} \qquad (4\text{-}27)$$

and the $|I_{ng}^2|$

$$\left|i_{ng}^2\right| = 4kT\beta g_g \Delta f \qquad (4\text{-}28)$$

$$R_g = \frac{1}{g_g} \cdot \frac{1}{Q^2+1} = \frac{1}{g_g} \cdot \frac{g_g^2}{\omega^2 C_{gs}^2} = \frac{g_g}{\omega^2 C_{gs}^2}$$

$$= \frac{\omega^2 C_{gs}^2}{5 g_m} \cdot \frac{1}{\omega^2 C_{gs}^2} = \frac{1}{5 g_m} \qquad (4\text{-}29)$$

and the $|Vng^2|$ is

$$\left|v_{ng}^2\right| = 4kT\beta R_g \Delta f \qquad (4\text{-}30)$$

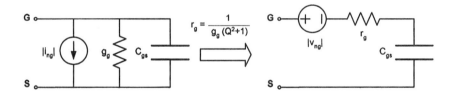

Figure 4-8 Converting Blue noise into white noise model

3.2   Calculating Noise figure in SPICE

Figure 4-9 shows how to simulate the noise figure using SPICE that is without the gate induced noise model. The conventional SPICE model only has a drain channel noise and the noise coefficient is (γ) is 2/3. However, in the short channel device, the γ is larger than 2/3, and the gate induced noise and the correlation noise must be added. The assumption of the simulation is that the M1 and M2 are dominant noise sources, so the induced gate noise and the correlation noise of M1 and M2 only are considered.

In this design, 0.35um CMOS technology is used with $\gamma= 1$ $\delta=2$, and $|c|=$ 0.395.

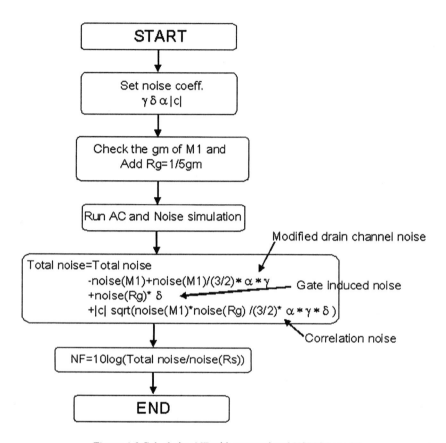

Figure 4-9 Calculating NF with conventional HSPICE model

3.3   Saving optimization time

In SPICE, 2 AC simulations are required to find the S parameters and the noise figure, and 1 transient simulation is needed to find IIP3. It usually takes a long time to run transient simulation. In order to save simulation time, first run AC simulations only and check the gain and noise figure; only if the gain and noise figure are better than specified values is a transient simulation run to find IIP3. If the gain and noise figure is worse than the specified values, a transient simulation is not run and a sub-optimum value of IIP3 is used (Fig. 4-10).

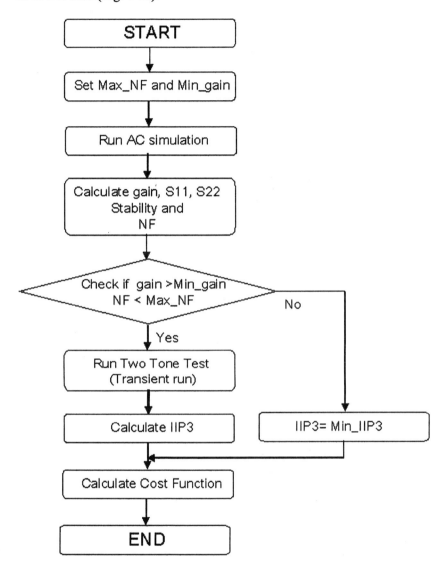

Figure 4-10 Saving optimization time by skipping IIP3 simulation

3.4   Cost function

To optimize a low noise amplifier, a cost function must first be specified. In this design, the cost function consists of 6 output variables: S_{11}, S_{22}, gain, IIP3, stability and noise figure. Table 4-1 shows the specifications for each of the outputs. Based on these specifications, the cost function is chosen as

$$
\begin{aligned}
Cost_{LNA} = {} & \frac{\frac{1}{3} - \frac{1}{NF}}{\frac{1}{3}} \times W_{NF} + \frac{6.5 - gain}{6.5} \times W_G \\[2ex]
& + \frac{1e-3-IIP3}{1e-3} \times W_{IIP3} + \frac{1 - stability}{stability} \times W_{st} \\[2ex]
& + \frac{\frac{1}{0.3} - \frac{1}{S11}}{\frac{1}{0.3}} \times W_{S11} + \frac{\frac{1}{0.3} - \frac{1}{S22}}{\frac{1}{0.3}} \times W_{S22}
\end{aligned}
\qquad (4\text{-}31)
$$

Gain, IIP3, S_{11} and S_{22} values are converted into linear values rather than dB values because the output in dB can take a negative value that is troublesome. If the output sign changes from positive to negative then the cost function is changed from positive to negative. This makes it difficult to find the optimum value if Eq. (4-31) is used. Since there are 6 weight functions required, a weighting coefficient of 17 is used for each weight in order to obtain a convenient maximum cost function value of 100.

3.5   Parameters to be optimized

The LNA circuit schematic is the same as that in Fig. 4-6, and the number of design parameters to be optimized is 8 including inductors, capacitors, bias point, load resistor, and transistor sizes.

3.6   Optimization Simulation Result

The essential advantages of parasitic-aware synthesis are illustrated in Table 4-1. With ideal inductors, all of the outputs easily meet the specifications. However, using parasitic-laden spiral inductors, most of the outputs can not meet the specification before optimization. Clearly, the parasitics of the passive components significantly degrade the performance of the low noise amplifier by effectively de-tuning it. Also, Table 4-1 shows the simulation results with parasitic-laden inductors after parasitic-aware optimization. As shown in Table 4-1, the use of CAD optimization significantly improves circuit performance; the final design including passive parasitics meets the original design specifications.

Table 4-1 Simulation result

	Specification	Ideal Inductor	parasitic laden inductor	
			without optimization	with optimization
NF	<3dB	1.94 dB	6.35 dB	3 dB
gain	16 dB	16.8 dB	4.7 dB	18 dB
IIP3	0 dbm	0 dBm	0 dBm	1.3 dBm
stability	>1	>1	>1	>1
S_{11}	<-10 dB	- 26 dB	-12 dB	-12 dB
S_{22}	<-10 dB	- 18 dB	- 3.5 dB	- 12 dB
I_{Bias}	5 mA	5 mA	5 mA	5 mA

Figure 4-11 shows the simulated noise figure vs. frequency. If the gate induced noise is not included the noise figure is about 2 dB. However, when the gate induced noise is considered, the noise figure increases to about 3 dB at 900 MHz. Hence, as shown in Fig. 4-11, the gate induced noise adds significant noise.

As previously mentioned, simulation time can be reduced by skipping transient simulations as shown in Table 4-2. It is quite clear that even when the total number of iterations is about same, the new saving algorithm simulated only 740 transient simulations and requires only 37 % of optimization time of the standard approach.

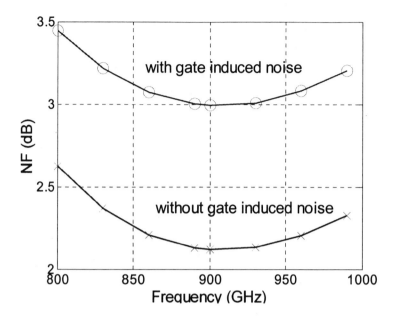

Figure 4-11 Noise Figure vs. Frequency

Table 4-2 Comparison of simulation time with and without IIP3 skipping algorithm (AC simulation time = 1.2 sec, Transient simulation time = 5 sec.)

	Total iteration with IIP3 skipping	Total iteration without IIP3 skipping
Total iteration	3087	3241
# of AC simulation	3087	3241
# of TR simulation	742	3241
Total simulation time	7414.4	20094.2
Normalized simulation time (%)	37	100

REFERENCES

[1] H. Nyquist, "Thermal agitation of electric charge in conductors," *Physics Review 32* pp. 110-113, 1928.

[2] R. Pettai, *Noise in Receiving Systems*, ch3, John Wiley & Sons, 1984.

[3] C.T. Sah, S. Y. Wu and F. H. Hielsher, "The effects of fixed bulk charge on the thermal noise in metal-oxide-semiconductor transistor," *IEEE Tran. On Electron Device*, vol 13, no 4, pp 410-414, April 1956.

[4] A. Van der Ziel, "Theory of shot noise in junction diodes and junction transistors," proc, IRE, pp. 1639-1646, Nov. 1955.

[5] H.T. Friis, "Noise figures of radio receivers," Proc. IRE, pp419-422, July 1944.

[6] T.H. Lee, *The design of CMOS Radio-Frequency Integrated circuits*, Cambridge University Press, ch. 11, 1998.

[7] B. Razavi, *RF Microelectronics*, Prentice Hall, Ch. 6, 1998.

[8] J. Zhou, "CMOS Low Noise Amplifier Design Utilizing Monolithic Transformers," *PHD Dissertation*, Oregon State University, 1998.

[9] G. Gonzalez, *Microwave Transistor Amplifiers Analysis and Design*, Prentice Hall 1996.

Chapter 5

OPTIMIZATION OF CMOS MIXERS

1. MIXER

The purpose of a mixer is to convert a base band or intermediate frequency to the higher RF frequency (up-conversion mixer) or to translate an RF frequency to a lower frequency base band or intermediate frequency (down-conversion mixer) by multiplying the input signal two signal with a Local Oscillator signal taken from the VCO in a frequency synthesizer. A basic architecture of an active mixer is the balanced mixer that is based on Gilbert multiplier concept [1]. When signals with two different frequencies are multiplied, the output frequencies consist of an additive term plus the subtractive term (Fig.5-1)

$$V_{out}(t) = \cos\omega_1 t \times \cos\omega_2 t$$
$$= \frac{1}{2}(\cos(\omega_1 - \omega_2)t + \cos(\omega_1 + \omega_2)t) \quad (5\text{-}1)$$

If the additive term (w_1+w_2) is used as output, it is called an up-conversion mixer. On the other hand, if the subtractive term (w_1-w_2) is used as the output, it is called a down-conversion mixer. An up-conversion mixer is typically used in the transmitter and a down-conversion mixer is often used in the receiver. The primary difference between down- and up-conversion mixers is in the filtering employed at output. That is, the filter of the up-conversion mixer removes the low-frequency difference term of Eq. (5-1) while the filter of the down-conversion mixer removes the additive term. Since the basic structures of those mixers are nearly identical, in this chapter, only the down-conversion mixer case will be discussed. The inputs to the down-conversion mixer are the RF signal and the (Local oscillator) LO signal that is mixed with it. The output signal is called the intermediate frequency (IF) signal, even though it is at base band frequencies in a homodyne architecture.

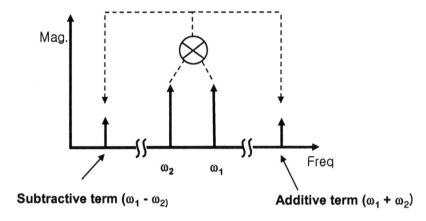

Figure 5-1 Multiplication of two different frequency terms

2.   SINGLE BALANCED MIXER

The basic schematic of a single-balanced mixer is shown in Fig. 5-2. As shown, the mixer consists of two parts. One is RF current generation by using M1, and the other is switching action by using M3 and M4. So the output will be a multiplication of the RF and the LO signals. The output frequency response of the mixer is shown in Fig. 5-2. If the LO signal is high enough, then it can be treated as a square wave because M3 and M4 work as switches. The output current equation is

$$i_{out}(t) = \text{sgn}[\cos\omega_{LO}t][I_{BIAS} + I_{RF}\cos\omega_{RF}t]$$

$$= \frac{4}{\pi}I_{BIAS}\cos\omega_{LO}t + \frac{4}{3\pi}I_{BIAS}\cos3\omega_{LO}t + \frac{5}{3\pi}I_{BIAS}\cos5\omega_{LO}t$$

$$+ \frac{2}{\pi}I_{RF}(\cos(\omega_{LO} - \omega_{RF})t + \cos(\omega_{LO} + \omega_{RF})t)$$

$$+ \frac{1}{\pi}I_{RF}(\cos(2\omega_{LO} - \omega_{RF})t + \cos(2\omega_{LO} + \omega_{RF})t)$$

$$(5-2)$$

where ω_{LO} is local oscillator frequency and ω_{RF} is RF frequency. As shown in Eq. (5-2), due to the bias current, the output current contains the ω_{LO} signal (Fig. 5-3). It is often troublesome to filter out the ω_{LO} signal and additive frequency term $(\omega_{LO} + \omega_{RF})$, if ω_{LO} and ω_{RF} are too close.

2.1   Conversion gain

One of the main specifications of a mixer is conversion gain. The two basic types are conversion voltage gain and conversion power gain. Conversion voltage gain is the gain from the RF input voltage to IF output voltage. So if the M3 and M4 are perfect switches, the conversion voltage gain of the single-balanced mixer is

$$G_V = \frac{2}{\pi}g_m \cdot R_{out}$$

$$(5-3)$$

Hence, the conversion power gain is

$$G_P = \left(\frac{2}{\pi}g_m \cdot R_{out}\right)^2 \frac{R_{in}}{R_{out}}$$

$$(5-4)$$

If R_{in} equals R_{out}, the conversion power gain is the same as the conversion voltage gain in dB. However, the output impedance R_{out} is usually bigger than R_{in}, so the power gain is smaller than the voltage gain. For example, if R_{out} is 1000 ohm and R_{in} is 50 ohm then the power gain is 13 dB (=10log(50/1000)) smaller than the voltage gain.

Other important specifications for an RF mixer are linearity, noise figure, and LO leakage.

2.2   Linearity

2.2.1   IIP3

As mentioned in Chapter 4, IIP3 and IIP2 represent the linearity of the RF circuit and the linearity o f t he last s tage i s the m ost i mportant for the overall system linearity performance. So, mixer linearity is important in determining the overall linearity. To measure the linearity, the two-tone test is used to obtain IIP3 as shown in Fig. 5-4.

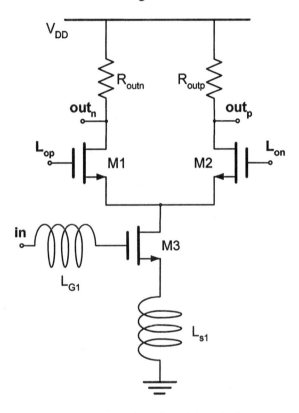

Figure 5-2 Schematic of single balanced mixer

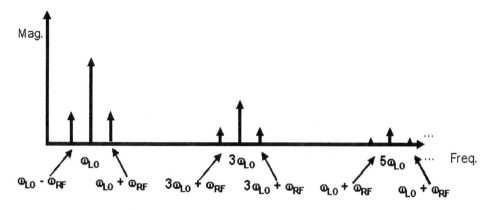

Figure 5-3 Output frequency spectrum of single balanced mixer

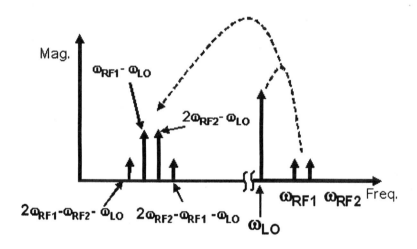

Figure 5-4 Frequency spectrum of two tone test of mixer

2.2.2   Calculating IIP3 and conversion power gain

As mentioned in Chapter 4, to calculate IIP3, the two-tone test is used. If the input versus output powers are plotted in dBm, the slope of the fundamental gain is one and the slope of the 3rd harmonic gain is three. According to the trigonometric ratio (Fig. 5-5), the IIP3 is

$$P_{IIP3} = C + \frac{A-B}{2}$$ (5-5)

Also gain can be found by subtracting the input power in dBm from the output power in dBm. So the power gain G is

$$G = A - C$$ (5-6)

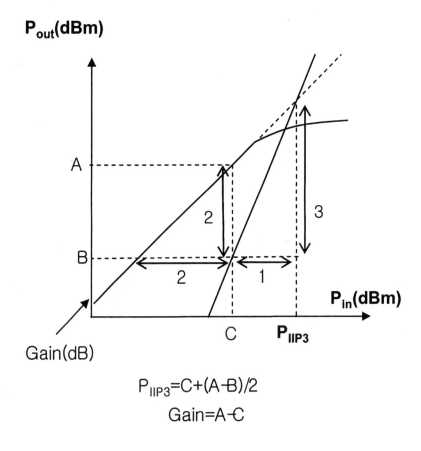

Figure 5-5 Calculating IIP3 and gain by using trigonometric ratio

2.3   Noise figure

2.3.1   DSB and SSB

Unlike the LNA, the mixer IF frequency spectrum is different from the input RF frequency due to the multiplication of the RF and LO signals. When two signals are multiplied, there are two different frequency terms showing up. One is the additive frequency term and the other term is the subtractive frequency term shown in Eq. (5-2). On the other hand, the output frequency spectrum also consists of two terms if the LO frequency is fixed. Figure 5-6 shows the noise spectrum of IF frequency. The IF term consists of two terms; one is ω_{LO}-ω_{RF} which is desired frequency term and the other is ω_{IM} -ω_{LO}. The image frequency is ω_{IM} and the relation among ω_{IM} , ω_{LO} and ω_{RF} is

$$|\omega_{IM} - \omega_{LO}| = |\omega_{RF} - \omega_{LO}| \qquad (5\text{-}7)$$

If the image frequency term is considered, the amount of noise in the IF frequency spectrum will be doubled. If the noise figure is considered neglecting the image noise, it is called double side band (DSB) noise figure. On the other hand, if the image noise is considered, then it is called single side band (SSB). The noise figure of SSB is 3 dB higher than DSB noise figure.

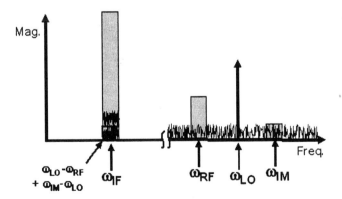

Figure 5-6 Accumulation of RF and image noise into the IF band

The frequency translation and time variance in mixers make it difficult to calculate the noise figure [2]. Since the input RF frequency and output frequency are different, the noise figure of a mixer cannot be calculated as it is done for an LNA. To find the noise of the mixer, the periodic steady-state (PSS) method is popular to find the noise of such a nonlinear system.

2.3.2   PSS simulation

The PSS method employs the shooting Newton method to find the periodic steady-state solution of the circuit. Since conventional SPICE does not have PSS analysis, it takes a long time to find the steady-state conditions of such a circuit if the solution is found at all. In order to find noise information of the circuit, frequency information is required. In order to find frequency information of the nonlinear circuit, the simulator needs to check if nodes of the circuit are in a steady-state condition. And if the nodes are in steady-state condition, the frequency information is then found by using the Fourier transform. The PSS simulation determines if nodes are in steady-state condition by using the Shooting Newton method [3],[4].

In order to find the noise figure, *SPECTRE* is used which has PSS analysis. The PSS analysis in *SPECTRE* is very useful to analyze RF circuits. PSS simulation can find most of RF specifications of nonlinear circuits, such as IIP3, 1 dB compression point, noise figure, large signal S-parameters, etc.

2.4   LO leakage

There are two types of LO leakage: one is the LO leakage to the RF input of the mixer that eventually reaches the antenna through reverse transmission through the LNA. The other is the LO leakage to the IF output of the mixer. The first case, LO leakage causes a self-mixing problem. That is, the LO leakage signal is mixed with the LO signal itself in the mixer and causes a significant DC offset problem. As mentioned before, if a single balanced mixer is used, the LO signal appears at the IF output, and the LO leakage signal makes the design of the IF filter more difficult because the LO leakage signal is close to IF frequency. Moreover, it desensitizes the following IF amplification stage. Most of the LO leakage problems are caused by transistor mismatches. The LO leakage problem is solved by employing a double-balanced mixer.

3. DOUBLE BALANCED MIXER

As mentioned before, the LO signal appears in as a component of the IF output spectrum in single-balanced mixers due to the bias current. To remove the LO signal from the output, double-balanced mixers are used as shown in Fig. 5-7. In the double-balanced mixer, another Gilbert multiplier cell is a dded c ompared to the single-balanced v ersion so that the DC bias current term at the output is cancelled. Hence, the LO signal does not appear at the IF output if the transistors are perfectly matched (Fig. 5-8).

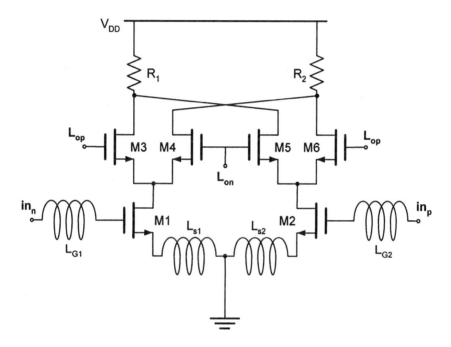

Figure 5-7 Schematic of double-balanced mixer

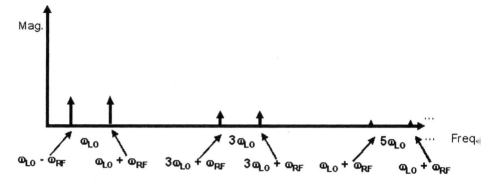

Figure 5-8 Output frequency spectrum of double balanced mixer

4.   DESIGN OF MIXERS

Typical specifications for a double-balanced mixer are as follows:

LO frequency: 828 MHz
RF frequency: 900MHz
IF frequency: 72 MHz
Conversion power gain: 3 dB
Conversion voltage gain: 16 dB
IIP3 >9 dBm
NF < 10 dB
BIAS current = 4 mA

To achieve the specifications of the mixer, a double-balanced configuration is used. Its input impedance can be matched to the desired 50 ohm standard by using source degeneration inductors without adding any addition thermal noise. The desired local oscillation frequency is 828 MHz, the input RF frequency is 900MHz; hence, the intermediate frequency (IF) is 72 MHz which typifies specifications for a GSM receiver.

As mentioned before, the mixer requires the use of a special simulation tool called periodic steady-state analysis to calculate noise for nonlinear switching circuit. Since conventional *SPICE* does not have PSS simulation capabilities, the *SPECTRE* simulator is used to design and optimize the mixer circuit. The sizes of transistors M1 and M2 are chosen based on the required conversion gain and the maximum allowable bias current. The size of switch devices M3, M4, M5, and M6 are decided based on the desired noise figure, linearity and switch loss. If the switching transistors are too big, they become too slow. On the other hand, if the switches are too small, low conversion gain and low linearity are obtained.

5. OPTIMIZATION OF MIXERS

In order to simulate the overall performance of the mixer, two different simulations are required. One is a PSS simulation to find the noise figure, and the other is a transient simulation to find IIP3 and the conversion power gain. As is the case in LNA optimization, total simulation time is significantly reduced by occasionally skipping expensive PSS simulations as shown in Fig. 5-9.

5.1 Cost function

The cost function consists of three outputs: conversion power gain, IIP3 performance, and noise figure.

$$
Cost_{Mixer} = \frac{\frac{1}{8} - \frac{1}{NF}}{\frac{1}{8}} \times W_{NF} + \frac{4 - gain}{4} \times W_G
$$
$$
+ \frac{1e - 3 - IIP3}{1e - 3} \times W_{IIP3} \tag{5-8}
$$

As in the LNA cost function case, the gain and IIP3 values are converted into equivalent linear values rather than maintained in dB form because the dB value may become negative. If the sign of the output changes from positive to negative it is difficult to find the optimum solution. There are three weight functions in Eq. (5-8) with each set to 33 to obtain a convenient maximum cost of 100.

5.2   Parameter to be optimized

The number of design parameters to be optimized in the double-balanced mixer is seven including inductors, transistors, capacitors, and resistors as shown in Fig. 5-7.

5.3   Optimization simulation results

Figures 5-10 to 5-12 show the NF, gain, and IIP3 distribution of the accepted solutions. More specifically, Figure 5-10 shows IIP3 versus gain, Fig. 5-11 shows NF versus gain, and Fig 5-12 shows NF versus IIP3; all three graphs are for the accepted solutions; i.e., those that meet or exceed the cost function objective. A circle on each plot identifies the optimum solution point. According to the Figs. 5-10 to Fig. 5-12, performance at the optimum solution point is

IIP3 = 9.2 dB
Conversion Power Gain = 3.5 dB
Noise Figure = 9.5 dB
BIAS current = 4 mA

So, all of the specifications of the mixer are satisfied after optimization.

As shown in Fig.5-10, there is trade off between IIP3 and gain. For example, if IIP3 is 9 dB, the gain is about 7 dB. On the other hand, if the gain is 10 dB, IIP3 is about only about 0 dB. Also, Fig.5-12 shows the trade off between noise figure and IIP3. For example, if IIP3 is 9 dB then noise figure is about 9.5 dB. On the other hand, if the noise figure is 5 dB then IIP3 is about 0 dB.

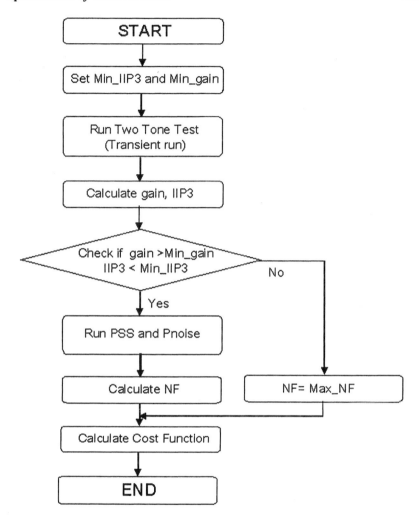

Figure 5-9 Saving optimization time by skipping IIP3 simulation

Figures 5-10 to 5-12 show that the bottleneck of the optimization is meeting the IIP3 specification. Fig. 5-11 shows that a lot of solutions satisfy the specifications of gain and noise figure. On the other hand, Fig. 5-10 and Fig. 5-12 show that only one solution satisfies the specifications because of the bottleneck in IIP3.

Optimization time is also saved by skipping PSS simulations as shown in Fig. 5-9; Table 5-1 shows a summary of the simulation time. As indicated, only about 84% of simulation time is required when using the PSS skipping algorithm

Table 5-1 Summary of the simulation time (Transient simulation time 30sec, PSS simulation time 10 sec)

	Total iteration with PSS skipping	Total iteration without PSS skipping
Total iteration	2119	2215
# of TR simulation	2119	2215
# of PSS simulation	1072	2215
total simulation time	74290	88600
Normalized simulation time (%)	84%	100

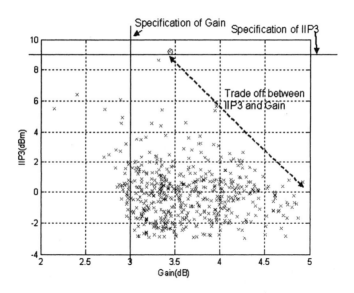

Figure 5-10 IIP3 versus conversion power gain for accepted solutions. (Circle indicates optimum solution point)

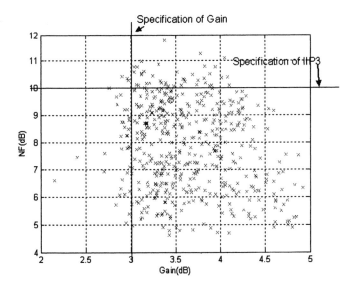

Figure 5-11 Noise figure versus conversion power gain of accepted solutions. (Circle shows optimum point)

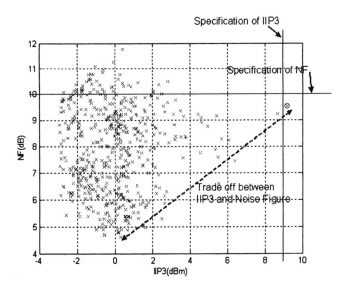

Figure 5-12 Noise figure versus IIP3 of accepted solutions. (Circle shows optimum solution point)

REFERENCES

[1] B. Gilbert "A Precise Four Quadrant Multiplier with subnosecond response,"
     IEEE J. Solid-State Circuits, pp. 365-373 December 1968.

[2] C.D Hull and R.G. Meyer, "A systematic Approach to the analysis of noise in
     Mixers," *IEEE Trans. Circuits and Systems* I, v. 40, no. 12, pp. 909-919. Dec.
     1993.

[3] http://www.cadence.com/datasheets/spectrerf.html, *RF Simulator (SpectreRF)
     User Guide.*

[4] Joel R. Phillips, "Analyzing Time-varying Noise Properties with SpectreRF,"
     *Affirma RF simulator (Spectre RF) user guide appendix I, Cadence Openbook IC
     product documentation,* Cadence Design System, 1998.

[5] T.H. Lee, *The design of CMOS Radio-Frequency Integrated circuits*, Cambridge
     University Press, 1998.

[6] B. Razavi, *RF Microelectronics*, Prentice Hall, 1998.

Chapter 6

OPTIMIZATION OF CMOS OSCILLATORS

Voltage controlled oscillators (VCOs) are one of the key building blocks in radio frequency communication systems and are used as local oscillators to translate data between baseband and frequencies suitable for wireless transmission. VCOs provide the LO signal to one of the ports of a mixer in a typical transceiver architecture as shown in Fig. 6-1. Since the quality of the modulated or demodulated signals is directly influenced by the spectral purity of the VCO, the phase noise of VCOs is one of the most critical performance parameters. Thus, t his chapter deals w ith the o ptimization of VCOs mainly focusing on the phase noise, which is preceded by basic oscillator design and study of phase noise models.

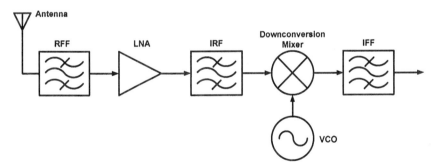

Figure 6-1 Heterodyne receiver architecture

1.   CMOS OSCILLATORS

Most oscillators employed in RF applications use LC resonators as shown in Fig. 6-2 (a); they are known to provide higher spectral purity and lower phase noise than other types such as ring oscillators, etc. [2]. Since the LC tank network is composed only of reactive components, the oscillating signal ideally maintains its oscillation amplitude without attenuation. The energy in the LC resonator transfers back and forth between the inductor and the capacitor in the form of the magnetic a nd e lectric energy without loss due to power dissipation. In practice, however, the quality factors of the inductor and the capacitor are finite. As a result, this leads to a practical parallel RLC network as shown in Fig. 6-2 (b). The quality factor Q of this network is defined as:

$$Q = \frac{R}{\sqrt{L/C}} = \frac{R}{\omega_0 L} = R \cdot \omega_0 C \qquad (6\text{-}1)$$

where ω_0 is the resonant frequency:

$$\omega_0 = \frac{1}{\sqrt{LC}} \qquad (6\text{-}2)$$

Intuitively, we note that the Q of this network is proportional to the value of the parallel resistor and infinite resistance leads to the ideal case in Fig. 6-2 (a). When the reactive components exchange their energies in this case, the amplitude of oscillating signal is damped over time due to the energy loss associated with the resistor [3]. So in order to sustain the oscillation, a practical tank network also needs an active circuit that provides a negative resistance to cancel out the positive loss resistance of the tank as shown in Fig. 6-2(c).

Since the active circuit must provide sufficient negative resistance in parallel to initiate oscillation, the start-up condition is:

$$R_{\text{tank}} \geq \left| -R_{\text{added}} \right| \qquad (6\text{-}3)$$

where R_{tank} and $-R_{\text{added}}$ are the resistances due to the tank and the negative resistance added from active devices, respectively. Note that the magnitude of the added negative resistance should be less than that of tank resistance because they are connected in parallel.

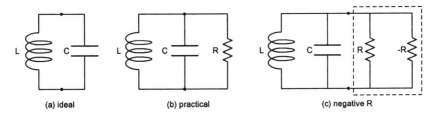

(a) ideal   (b) practical   (c) negative R

Figure 6-2 LC resonators

One way to provide negative resistance or conductance is by using a pair of cross-coupled transistors as shown in Fig. 6-3(a). A small-signal counterpart is also shown in Fig. 6-3(b). In order to find the impedance looking into the cross-coupled devices, we simplified the problem by neglecting the r_{ds} of the transistor and applying the test voltage source at one input.

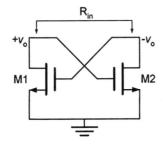

(a) cross-coupled NMOS pair

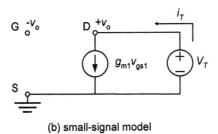

(b) small-signal model

Figure 6-3 Negative-Gm generation using a cross-coupled pair

Note that the voltages at the gate and the drain are exactly $180°$ out of phase. From this small-signal model, the test current is equal to:

$$i_T = g_{m1} v_{gs1} = g_{m1} \cdot (-v_o) = g_{m1} \cdot (-v_T) \qquad (6\text{-}4)$$

and the input resistance is:

$$\frac{v_T}{i_T} = -\frac{1}{g_{m1}} \qquad (6\text{-}5)$$

Since the calculated impedance only accounts for one side of the cross-coupled device, the total input impedance in a differential configuration is twice that value or $-2/g_m$.

As mentioned earlier in this chapter, VCOs in RF applications should drive the mixer, most of which use a fully differential topology. For that reason, we deal only with differential output VCOs in this chapter. There are a couple of advantages for using a differential topology over a single-ended one. First, the differential VCOs are insensitive to common-mode noise components. Since most of extrinsic noise effects such as substrate and supply noise amplification are common-mode in nature, the differential topology shows a good overall noise performance. Secondly, the differential excitation can boost the Q of the inductor in the tank, which is usually the main source of the problem in VCO design, by reducing the effects of shunt parasitics of on-chip inductors as described in Chapter 2 [4].

Tuning range and phase noise are very important performance measures in VCO design. The latter will be extensively discussed in the following section. In most transceiver architectures, the front-end up- and down-conversion functions select one of channels as seen previously in Fig. 6-1. This channel selection is accomplished by adjusting the resonant frequency of the tank by through the controlling voltage of the VCO used in the frequency synthesizer. As studied in Chapter 2, voltage-controlled varactors are often used to tune the oscillation frequency.

Ring oscillators shown in Fig. 6-4 realize a very large tuning range by allowing control of the bias current. However, the use of ring oscillators is limited due to the poor phase noise performance, low operating frequency, and high power consumption, despite the fact that it is very simple topology that is easily integrated with other digital circuits [5].

On the other hand, the typical VCO structure incorporating the voltage-controlled varactor is shown in Fig. 6-5. In this topology, the tuning range is determined by the variable capacitance range of the varactor relative to the total fixed capacitance associated with nodes A and B including all parasitic capacitances. The fixed capacitance must be minimized to allow variations in the varactor capacitance to have a larger effect on the resonant frequency.

Even w ith s mall l oads, t he t uning r ange i s typically l imited t o a bout 1 5% [6]-[8].

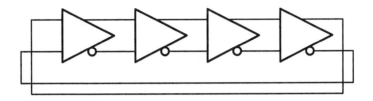

Figure 6-4 Differential four-stage ring oscillator

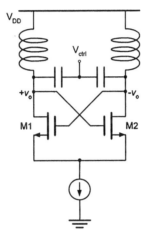

Figure 6-5 Differential CMOS VCO using a varactor for tuning

2. PHASE NOISE

Phase noise is one of the most critical measures that decides the overall performance of VCOs. This section analyzes the effects of the presence of phase noise on transceivers and deals with two alternative phase noise models: the Leeson model and the Hajimiri model.

2.1 Effects of phase noise

A general oscillator signal can be expressed as:

$$V(t) = A(t)\cos(\omega_c t + \varphi(t))$$ (6-6)

where $A(t)$ is the amplitude and $\varphi(t)$ is a small random excess phase representing variations of the oscillating signal. For an ideal oscillator, $A(t)$ and $\varphi(t)$ are both constants and the output spectrum is an impulse at frequency ω_c. However, in practice, $A(t)$ and $\varphi(t)$ are functions of time, which leads to a non-ideal spectrum response shown in Fig. 6-6. This spectrum response is referred to as exhibiting *sidebands* or *noise skirts* because of its shape.

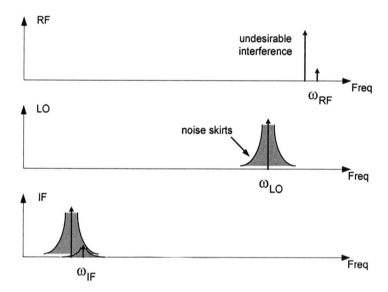

Figure 6-6 An example of harmful effects of phase noise

Note that the desired signal can be buried under the phase noise skirts of an adjacent strong channel, reducing the effective dynamic range of the receiver.

Phase noise is quantified in terms of the single sideband noise spectral density in the frequency domain as [9]:

$$L(\Delta\omega) = 10\log\left[\frac{P_{sideband}\left(\omega_c + \Delta\omega, 1Hz\right)}{P_{carrier}}\right] \tag{6-7}$$

where $P_{sideband}(\omega_c+\Delta\omega, 1Hz)$ is the single sideband power at an offset frequency $\Delta\omega$ with respect to the carrier frequency ω_c in a unit bandwidth, and $P_{carrier}$ is the carrier power. Therefore, the unit for phase noise by definition is dBc/Hz, which reads decibel with respect to carrier per Hertz (Fig. 6-7).

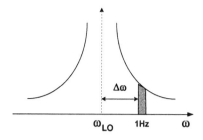

Figure 6-7 Phase noise with respect to carrier power per unit bandwidth

2.2 Leeson phase noise model

In 1966, Leeson published a derivation of phase noise in terms of known oscillator parameters [10]. The model is based on a linear time invariant approach for tuned tank oscillators. The phase noise of an oscillator is predicted using the following equation:

$$L\{\Delta\omega\} = 10\cdot\log\left\{\frac{2FkT}{P_{sig}}\cdot\left[1+\left(\frac{\omega_0}{2Q_L\cdot\Delta\omega}\right)^2\right]\cdot\left(1+\frac{\omega_{1/f^3}}{|\Delta\omega|}\right)\right\} \tag{6-8}$$

where F is the device noise excess factor or simply noise factor, which is a fitting parameter derived from measured data, k is Boltzmann's constant, T is the absolute temperature, P_{sig} is the average power dissipated in the resistive part of the tank, ω_0 is the oscillation frequency, Q_L is the loaded quality factor of the tank, and $\omega_{1/f}$ is the frequency of the corner between the $1/f^3$ region and $1/f^2$ regions, as shown in Fig. 6-8. At a glance of Leeson's model formula, an oscillator with a large Q of the tank results in better phase noise immunity. In addition to the Q of the LC tank, active devices contribute a significant noise to the oscillator. Unfortunately, there is no known way to directly calculate F from all noise sources. Leeson's phase noise model gives some design insight, but it is not properly predicting the effect of time-varying noise sources.

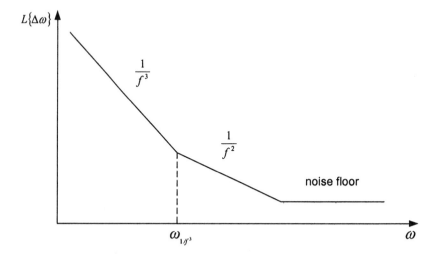

Figure 6-8 Typical phase noise spectrum

2.3   Hajimiri phase noise model

The time-variant phase noise model of an oscillator was introduced by Hajimiri [11]. The main idea of Hajimiri's phase noise model is to characterize the phase noise of an oscillator by its phase impulse response. It implies that a given voltage or current impulse should cause phase shift and phase noise can be characterized by an impulse response. The phase noise predicted using Hajimiri's model is:

$$L\{\Delta\omega\} = \frac{\Gamma_{rms}^2 \cdot i_n^2 / \Delta\omega}{16\pi^2 \omega_{off}^2 \cdot q_{max}^2} \tag{6-9}$$

where Γ_{rms} is the root mean square value of the impulse sensitivity function (ISF), $i_n^2/\Delta\omega$ is the power spectral density of a white input noise source, q_{max} is the maximum charge across the capacitor of the tank.

In Hajimiri's formula, ISF is a function that shows the sensitivity of every point on the waveform to an input impulse. When a given perturbation causes a large phase shift, ISF is large, while it is small in the opposite case [9]. It is zero at the peak and maximum at zero crossing in the case of an LC oscillator. The phase impulse response of an ideal tank is shown in Fig. 6-9. The time-variant phase noise model can predict phase noise, but it is difficult to define the ISF function for simulation.

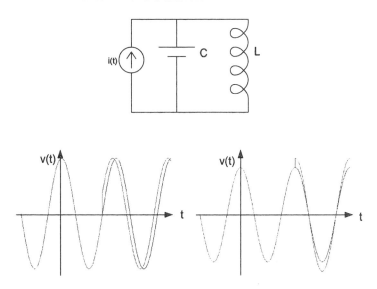

Figure 6-9 Phase impulse responses of an ideal LC oscillator

3.   DESIGN OF VCO

The complementary cross-coupled LC VCO in Fig. 6-10 is selected to demonstrate the parasitic-aware design technique VCOs. This design has several advantages over the VCO structure of Fig. 6-5. First, it re-uses the current with an additional pMOS pair and hence provides extra g_m. For that reason it needs only a half the current of nMOS-only structure for the same total g_m. Secondly, the oscillation amplitude of this structure is approximately twice as large than that of nMOS-only structure. These properties result in a better phase noise performance for a given tail current. However, use of the complementary structure may decrease the tuning range due to extra transistor parasitic capacitances.

Table 6-1 summarizes the design specifications. From Fig. 6-10, the tank parameters are:

$$f_c = \cfrac{1}{2\pi\sqrt{\cfrac{L}{2}\cdot C}}$$

$$(6\text{-}10)$$

$$LC = \cfrac{1}{2\pi^2 f_c^2}$$

where f_c is the target oscillation frequency.

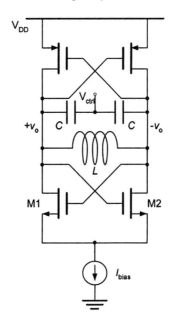

Figure 6-10 Complementary cross-coupled LC tuned VCO

The Q of the tank is commonly dominated by the low Q of on-chip inductors, which rarely exceeds 10. Assuming the Q of the tank is between 3~10, R_{tank} can be estimated. To satisfy the start-up condition in Eq. (6-3), the input resistance from Eq. (6-4) should be

$$\frac{2}{g_{mn} + g_{mp}} \le R_{tank}$$

$$(6\text{-}11)$$

Assuming $g_{mn} = g_{mp}$,

$$\frac{2}{2g_{mn}} \leq R_{tank}$$

$$\frac{1}{R_{tank}} \leq g_{mn} = k_n' \left(\frac{W}{L}\right)_n (V_{gs} - V_T)$$ (6-12)

With a fixed bias current,

$$\frac{I_{bias}}{2} = \frac{k_n'}{2} \left(\frac{W}{L}\right)_n (V_{gs} - V_T)^2$$ (6-13)

Combining Eq. (6-13) and Eq. (6-14), the possible solution can be located as

$$\left(\frac{W}{L}\right)_n > \frac{1}{I_{bias} k_n' R_{tank}^2}$$ (6-14)

$$(V_{gs} - V_T) < I_{bias} R_{tank}$$

Table 6-1 Summary of design specs

Parameters	
Process	CMOS 0.18-μm 1.8-V single power supply
Oscillation frequency	2.4GHz
Tuning range	> 15%
Phase noise @1MHz	<-124 dBc/Hz
I_{tail}	2 mA
Number of design variables	5
Number of objectives	3

From Eq. (6-13) and Eq.(6-14), the plot of Fig. 6-11 shows the design regime satisfying the start-up.

4. OPTIMIZATION OF CMOS VCO

As we discussed in the previous sections, phase noise is a very important parameter that decides the overall performance of VCOs. In order to compare VCOs operating at different frequencies, a figure of merit (FOM) is widely used which is defined as

$$FOM = 10 \cdot \log_{10}\left[\left(\frac{\omega_0}{\Delta\omega}\right)^2 \cdot \frac{1}{L(\Delta\omega)\cdot P}\right] \tag{6-15}$$

where P is the average power consumption of the oscillator. Note that FOM accounts for phase noise if the power dissipation is fixed but it does not include the tuning range, which is another important performance factor in VCOs. In this section, the tuning range as well as the phase noise are optimized using the parasitic-aware design technique.

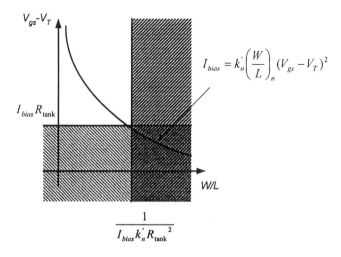

Figure 6-11 Design regime for start-up condition of oscillation

4.1   Optimization of VCO

From Fig. 6-11, we obtain the minimum sizes of the transistors needed for a successful start-up. However, it is extremely difficult to find the optimum design by minimizing the phase noise and maximizing the tuning range simultaneously. For example, to maximize the tuning range, it is necessary to minimize the parasitic capacitances and hence reduce the size of transistors. If the size becomes too small, however, the flicker noise can dominate, which in turn would increase the phase noise. The complexity of the problem gets even worse when considering the parasitics of inductors, which exhibit nonlinear functions of inductance and frequency.

For parasitic-aware design and optimization of the VCO, *SPECTRE* PSS/Pnoise simulation and particle swarm algorithm are used. Figure 6-12 illustrates one optimization loop; this loop repeats until the predefined maximum number of iterations is reached. If running PSS/Pnoise simulation with the synthesized *netlist*, it generates output files, which include all noise values and the oscillation frequency. Basically, the oscillation frequency should be located close to the target oscillation frequency. Thus, to avoid wasting unnecessary simulation time, it screens out the circuits of which oscillation frequencies are far away from the target 2.4GHz (higher than 2.5GHz and lower than 2.3GHz in this example) without further advancing to the tuning range simulations. In this case it uses a cost function that constitutes only the oscillation frequency. After extracting appropriate data such as oscillation frequency, carrier power, and output noise values at different frequency offsets, etc., from the output file, the phase noise is obtained by subtracting the carrier power from the output noise at the specified offset frequency.

To find out the tuning range of the target circuit, it is necessary to sweep the varactor control voltage with at least two different values, V_{ctrl_min} and V_{ctrl_max}. Then the tuning range is

$$f_c(V_{ctrl\_max}) - f_c(V_{ctrl\_min}) \qquad (6\text{-}16)$$

where f_c is the oscillation frequency which is the function of control voltage.

4.2 Optimization results

Table 6-2 Summary of optimization results

Parameters	Before optimization	After optimization
Oscillation frequency	2.4 GHz	2.4 GHz
Phase noise (dBc/Hz) @1MHz	-122.5	-127.0
Tuning range	±14%	±17%
I_{tail}	2 mA	2 mA
Output amplitude	0.84 V	0.97 V

Table 6-2 summarizes the VCO's performance achieved after optimization. There are noticeable differences between the un-optimized VCO and the synthesized VCO. The phase noise has been improved by 4.5dB, the tuning range by 3%, and the output swing by 15% through the parasitic-aware optimization process. Note that the output swing amplitude has also been improved even though it was not considered as an optimization objective. The reason is that the optimizer increased the Q of the tank for phase noise, which in turn would increase R_{tank} and the output amplitude, which is proportional to R_{tank} and I_{bias}.

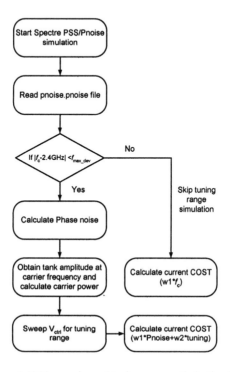

Figure 6-12 Phase noise and tuning range optimization loop

Figure 6-13 illustrates vividly how effectively the parasitic-aware design technique optimizes the phase noise and the tuning range. Both phase noise and tuning range improve dramatically as the total iterations increase. Notice that there is a jump of the tuning range at around 13 iteration but it steps down a little again by trading the tuning range for the phase noise. Figure 6-14 shows this trade-off more clearly. The trend trajectory is pointing towards the upper-left corner. This trend is desirable since it indicates the optimizer is appropriately doing its job by trying to find solutions with wider tuning range and smaller phase noise. It eventually hits a boundary called the *trade-off line* as the constraints become tighter. From this point, the optimization can be performed in two directions, down-left or upper-right, depending on the weight for each objective in cost function. In this example the phase noise has been weighted more than the tuning range and we can notice that acceptable solutions are more crowded in lower-left corner.

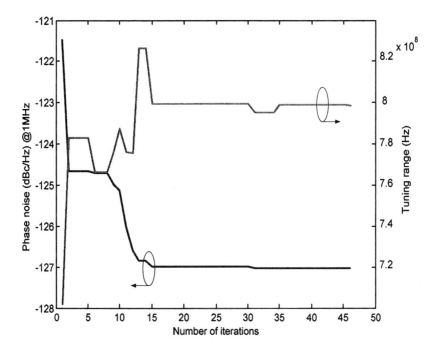

Figure 6-13 Phase noise and tuning range vs. number of iterations

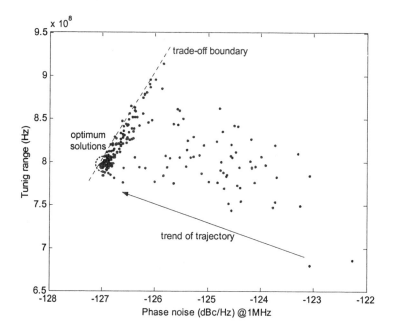

Figure 6-14 Tuning range vs. phase noise

REFERENCES

[1] Min Chu, "A fully integrated quadrature LC oscillator for applications in the unlicensed national information-infrastructure frequency band," *M.S. thesis*, University of Washington, Seattle, WA, Nov. 2002.

[2] B. D. Muer, M. Borremans, M. Steyaert, and G. L. Puma, "A 2-GHz low-phase-noise integrated LC-VCO set with flicker-noise upconversion minimization," *IEEE J. Solid-State Circuits*, vol. 35, pp. 1034-1038, July 2000.

[3] D. Ham and A. Hajimiri, "Concepts and methods in optimization of integrated LC VCOs," *IEEE J. Solid-State Circuits*, vol. 36, pp. 896-909, June 2001.

[4] S. Kodali and D. J. Allstot, "A symmetric miniature 3D inductor," *to be published at ISCAS* 2003.

[5] T. H. Lee and A. Hajimiri, "Oscillator phase noise: a tutorial," *IEEE J. Solid-State Circuits*, Vol. 35, No. 3, pp. 326-336, Mar. 2000.

[6] P. Andreani, "A low-phase-noise low-phase-error 1.8GHz quadrature CMOS VCO," *Digest of Technical Papers, IEEE International Solid-State Circuits Conference*, Vol. 1, pp. 290-291, 2002.

[7] A. Rofougaran, J. J. Rael, M. Rofougaran, and A. A. Abidi, "A 900Mhz CMOS LC-oscillator with quadrature outputs," *Digest of Technical Papers, IEEE International Solid-State Circuits Conference*, pp. 392-393, 1996.

[8] M. Tiebout, "Low-power, low-phase noise differentially tuned quadrature VCO design in standard CMOS," *IEEE J. Solid-State Circuits*, vol. 36, pp. 1018-1024, July 2001.

[9] A. Hajimiri and T. H. Lee, *The Design of Low Noise Oscillator*, Kluwer Academic Publishers, Feb. 1999.

[10] E. D. B. Leeson, "A simple model of feedback oscillators noise spectrum," *Proceedings of the IEEE*, Vol. 54, pp. 329-330, 1966.

[11] A. Hajimiri and T. H. Lee, "A general theory of phase noise in electrical oscillatos," *IEEE J. Solid-State Circuits*, vol. 33, pp. 179-194, Feb. 1998.

Chapter 7

OPTIMIZATION OF CMOS RF POWER AMPLIFIERS

1.   RF POWER AMPLIFIERS

The main purpose of the PA is to amplify the transmit signal so that it can reach the receiver at a specified distance. Depending on the distance between transmitter and receiver, the output power of PA is determined, and depending on the method of modulation, the type of power amplifier is decided. If the modulation scheme used is highly linear such as Binary Phase Shift Keying (BPSK), a linear PA is required. Since the PA dissipates a lot of the power, high efficiency is necessary to improve the battery life for mobile wireless systems. Also if the power efficiency is low, the temperature of the chip will be increased which degrades the performance and an expensive and bulky cooling system may be required. So, the power efficiency and linearity of the PA are very important. Usually, a linear PA has lower power efficiency than a nonlinear PA. There are several types of PAs, class-A, B, AB, C, D, E, and F. Class-A, B, AB are highly linear PAs, while class-C is somewhat nonlinear PA. Class-D, and Class-E are high-efficiency nonlinear switching PAs. Class-F is a high efficiency nonlinear PA obtained by creating a square-wave overdrive voltage using harmonic shaping. First, Class-A, B, AB, C amplifiers are discussed.

1.1   Linear Amplifier: Class-A, B, AB, and C

Figure 7-1 shows a basic circuit of a linear PA where in the transistor is used as a current source [1]-[2]. The transistor changes the RF input voltage into an output drain current. The RF choke inductor should be large enough to block passage of the RF signal to V_{DD}. Class-A, class-B, class-AB and class-C operation is determined by the input DC bias voltage, which sets the conduction angle. The conduction angle represents the fraction that a transistor is on during the RF cycle. If the transistor is on during the entire RF cycle, the conduction angle is 360 degrees, and if it is on during half the RF cycle, the conduction angle is 180 degrees. Drain efficiency, total harmonic distortion (THD), output power, and power gain depend on the conduction angle.

There are two key matching networks in Fig. 7-1; one is the input matching network and the other is output matching network. The input matching network is usually designed based on the maximum power transfer theory, and the output matching network is designed to achieve the specified output power. For example, if the power amplifier requires 1 W output power, and supply voltage is 3 volts, then the effective resistance (R_{EFF}) is only about 5 ohm. So, the output matching network is designed to interface the 5 ohm impedance seen by the driver to the 50 ohm antenna impedance.

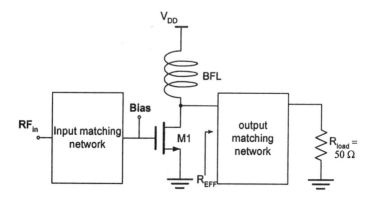

Figure 7-1 Simplified circuit of linear power amplifier

1.1.1   Class-A Power Amplifier

As mentioned before, Class-A, B, C, and AB operation is decided by the conduction angle. Figure 7-2 shows the normalized signals for a class-A power a mplifier. S ince t he t ransistor i s o n d uring t he e ntire R F cycle, t he conduction observed is 360 degrees, and since the transistor is always on, it is a very linear power amplifier as shown.

The definition of drain efficiency η is

$$\eta = (\frac{P_{RFout}}{P_{DC}} \cdot 100) \tag{7-1}$$

Since drain efficiency η depends only on the output power drawn from DC power supply, it does not show the relation between output power and input power. Therefore, there is another method to represent the efficiency; power added efficiency (PAE) is defined as

$$PAE = (\frac{P_{RFout} - P_{RFin}}{P_{DC}} \cdot 100) \tag{7-2}$$

Ideally, the maximum value of V_{ds} is V_{DD}. So, the output power is

$$P_{RFout} = \frac{V_{DD}^{2}}{2R_{EFF}} \tag{7-3}$$

where R_{opt} is the resistance required for this voltage output level. The power drawn from the supply is

$$P_{DC} = \frac{V_{DD}^{2}}{R_{EFF}} \tag{7-4}$$

So, the maximum drain efficiency is

$$\eta = (\frac{P_{RFout}}{P_{DC}} \cdot 100) = \frac{\dfrac{V_{DD}^{2}}{2R_{EFF}}}{\dfrac{V_{DD}^{2}}{R_{EFF}}} \cdot 100 = 50\% \tag{7-5}$$

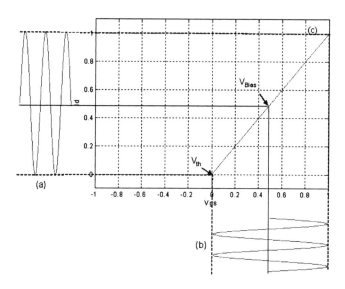

Figure 7-2 (a) Normalized transient response of output current (I_D), (b) Normalized input voltage (V_{GS}) and (c) I_D Vs V_{GS} with DC bias point for a class-A power amplifier

Since the maximum drain efficiency is 50%, the class-A drain efficiency cannot exceed that value. Practically, since the maximum V_{ds} value is less than V_{DD} a nd m atching n etwork is n ot l ossless, drain e fficiency i s always less than 50%.

1.1.2   Class-B Amplifier

If the conduction angle of the PA is 180 degrees, it operates in a class-B mode. Figure 7-3 shows the normalized signals for a class-B power amplifier.

As shown in Fig. 7-3, the transistor is on during only half of the RF cycle. Therefore, the drain current I_d is a half sine wave. Using the Fourier series, the DC and fundamental current components are found. If the amplitude of the half sine wave is A, the fundamental and DC components are given by

$$I_D \cong \frac{A}{\pi} + \frac{A}{2} \cdot \sin(\omega_1) \qquad (7\text{-}6)$$

So, the DC current component is

$$I_{DC} \cong \frac{A}{\pi} \qquad (7\text{-}7)$$

and the fundamental component is

$$I_{fund} \cong \frac{A}{2} \qquad (7\text{-}8)$$

The output power is

$$P_{RFout} = I_{fund}^2 \cdot R_{EFF} = \frac{A^2}{4} \cdot R_{EFF} \qquad (7\text{-}9)$$

and the DC power dissipation is

$$P_{DC} = V_{DD} \cdot I_{DC} = V_{DD} \cdot \frac{A}{\pi} \qquad (7\text{-}10)$$

Finally, the maximum drain efficiency of the class-B PA is

$$\eta = (\frac{P_{RFout}}{P_{DC}} \cdot 100) = \frac{\dfrac{A^2}{4} \cdot R_{EFF}}{V_{DD} \cdot \dfrac{A}{\pi}} \cdot 100 = \frac{\pi}{4} \cdot 100 = 78.54\% \qquad (7\text{-}11)$$

where $I_{fund} * R_{EFF} * 2 = VDD$

Maximum drain efficiency in class-B operation is 78.54%. However, since the transistor is not an ideal current source and the matching network is not a lossless network, drain efficiency is less than 78.54%, but significantly higher than for class-A operation. However, output power is lower than class-A, and linearity is worse because V_{ds} contains several harmonic components as well as a fundamental frequency.

1.1.3 Class-C Amplifier and Class-AB Amplifier

If the conduction angle of the PA is less than 180 degrees, operation is class-C; if the conduction angle is between 360 and 180 degrees, operation is class-AB. Figure 7-4 shows the normalized signals for a class-C power amplifier and Fig. 7-5 shows the same waveforms for a class-AB stage.

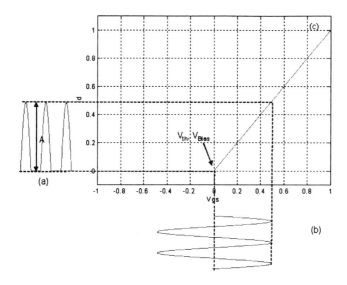

Figure 7-3 (a) Normalized transient response of output current (I_D), (b) Normalized input
voltage (V_{GS}) and (c) I_D Vs V_{GS} with DC bias point for a class-B power amplifier

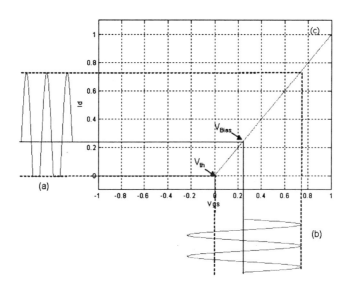

Figure 7-4 (a) Normalized transient response of output current (I_D), (b) Normalized input
voltage (V_{GS}) and (c) I_D Vs V_{GS} with DC bias point for a class-C power amplifier

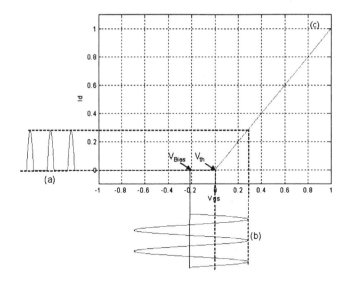

Figure 7-5 (a) Normalized transient response of output current (I_D), (b) Normalized input voltage (V_{GS}) and (c) I_D Vs V_{GS} with DC bias point for a class-AB power amplifier

As mentioned in the discussion of class-B amplifiers, reducing the conduction angle increases its drain efficiency, but reduces both linearity and output power. Therefore, at first glance, a class-C PA has high drain efficiency, low linearity, and low output power. On other the hand, the class-AB PA has low drain efficiency, high linearity, and high output power. To show the overall relation between conduction angle, efficiency, output power, l inearity, and n ormalized g ain, the o utput c urrent i s e xpanded i n a Fourier series. Using the *FFT* method, drain efficiency, normalized output power, and total harmonic distortion (THD) are easily calculated.

Figure 7-6 shows the drain efficiency versus conduction angle. As shown, when the conduction angle is 0 degrees, the maximum drain efficiency η is 100%; maximum drain efficiency of the class-B PA whose conduction angle is 180 degree is 78.5%.

Figure 7-7 shows the normalized output power versus conduction angle. When the conduction angle is 360 degree, the output power has its maximum value. The equation of the normalized RF output power is

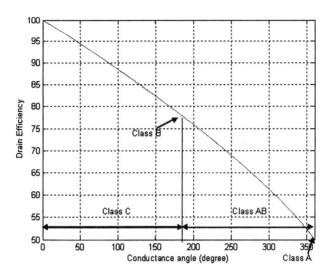

Figure 7-6 Drain Efficiency vs Conduction Angle for PA of Fig. 7-1

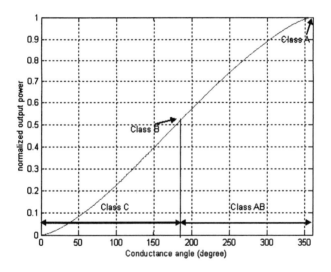

Figure 7-7 Normalized Pout vs Conduction Angle for PA of Fig. 7-1

Figure 7-8 shows total harmonic distortion (THD). As mentioned before, when conduction angle is 360 degree, THD is minimum, corresponding to highly linear operation. The equation of THD is

$$THD = \frac{\sum_{n=2}^{\infty} Id_n}{Id_1} \cdot 100 \qquad (7\text{-}12)$$

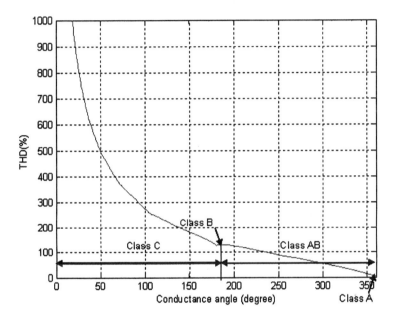

Figure 7-8 Total Harmonic Distortion over Fundamental versus Conduction Angle for the PA of Fig. 7-1

1.2 Nonlinear power amplifiers: Class-F and Class-E

As mentioned before, a nonlinear PA has higher drain efficiency than a linear amplifier. There are two methods to increase the efficiency of nonlinear PA. One method involves squaring the shape of the sine wave to achieve sharper turn on and turn off, and the other involves switching the power transistor between the OFF and ON states. The former is called a class-F PA, and the latter is a class-E PA [1]-[2].

1.2.1   Class-F PA

To re-shape a sine wave into a square wave, the third harmonic is very important. Figure 7-9 shows the third harmonic squaring effect on a sine wave. As shown, when $V_3=1/9 V_1$, the drain voltage waveform is maximally flat. When $V_3=1/6 V_1$, the wave has a minimum peak value. When $V_3=0.4 V_1$, the peak value is same as when $V_3=0$. Therefore, when V_3 is higher than $1/9 V_1$, the wave overshoots and its peak value is decreased until $V_3=1/6 V_1$. Above $V_3=1/6 V_1$, the peak value increases, and when $V_3=0.4 V_1$, the peak value is the same as for $V_3=0$. The merit of squaring sine the wave shape is as follows: When the peak value of the wave becomes lower by adding the third harmonic, the contribution ratio between the fundamental frequency value and the peak value is changed. Since the peak value becomes lower, the contribution ratio is increased so that the efficiency is increased by a factor of V_1/V_3.

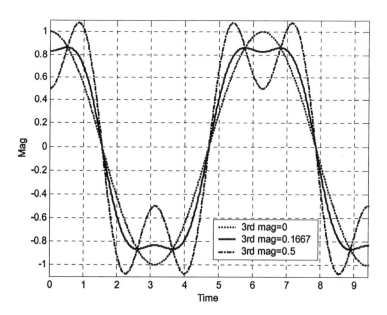

Figure 7-9 Third harmonic squaring effect on a sine wave

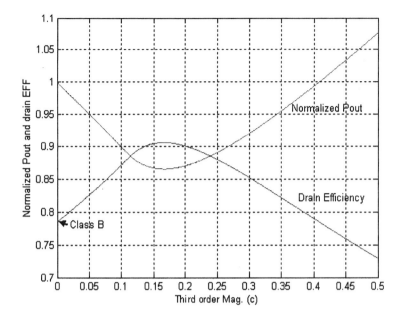

Figure 7-10 Peak value and Efficiency versus third harmonic magnitude in class-B PA

Figure 7-10 shows how drain efficiency is affected by varying the third harmonic. When $V_3 = 1/6\ V_1$, the peak voltage is $V_{pk} = 0.8660\ V_1$. If the input stage of class-F PA is same as for the class-B PA, whose maximum drain efficiency is 78.5%, then the drain efficiency of class-F will be

$$\eta = 78.5 \cdot \frac{V_1}{V_{pk}} = 78.5 \cdot \frac{1}{0.866} = 90.7\% \tag{7-13}$$

Figure 7-11 shows a general circuit diagram of a class-F-PA. It appears as a linear amplifier except the drain load shorts even harmonics in order to allow the third harmonic squaring effect on the drain voltage. Therefore, to increase the efficiency, a 0 to 0.4 third harmonic relative to the first harmonic needs to be added.

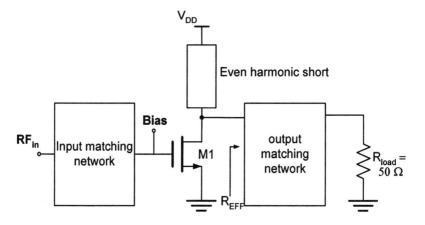

Figure 7-11 Class-F PA circuit diagram

1.2.2   Class-E PA

The other technique to increase drain efficiency η is the use of a switching PA. Figure 7-12 shows the general circuit diagram of class-E PA. It is assumed that transistor behaves as a switching component, rather than as a current source. How does the switching amplifier increase drain efficiency?

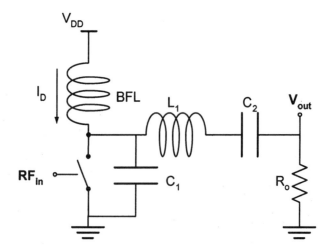

Figure 7-12 Class-E PA circuit diagram

Figure 7-13 shows the voltage and current waveform at the output and drain of the class-E PA for an ideal switch. When the drain current is maximum, the drain voltage is zero, and when the drain voltage is maximum, the drain current is zero. That is, the power, which is voltage times current, is zero. So, all the power drawn from DC source driven into the output node; that is efficiency is theoretically 100%. However, when the switch is off, leakage current still flows because of the non-ideal properties of the transistor. When does the class-E PA achieve its maximum drain efficiency? According to [3],[4], we can analyze the class-E PA as follows.

After long calculation, the RF output power is

$$P_{out} = 0.577 \cdot \frac{V_{DD}^2}{R_{opt}} \qquad (7\text{-}14)$$

where R_{opt} is optimum load output resistance in Fig. 7-12, and V_{DD} is power supply voltage.

$$L = \frac{Q \cdot R_{opt}}{\omega} \qquad (7\text{-}15)$$

Q is the quality factor of the output network of class-E in Fig. 7-12, and ω is the operating frequency.

$$C_1 = \frac{1}{\omega \cdot (R_{opt} \cdot 5.447)} \qquad (7\text{-}16)$$

$$C_2 = C_1 \cdot \frac{5.447}{Q} \cdot (1 + \frac{1.42}{Q - 2.08}) \qquad (7\text{-}17)$$

Using these equations, all components can be found. There are several advantages of the class-E amplifier [8]. First, a class-E PA requires a larger output impedance than other types of PAs. For example, to get 1 watt of output power, R_{opt} in Class-E PA occurs when $V_{DD}=3.3V$

$$R_{opt} = 0.577 \cdot \frac{V_{DD}^2}{P_{out}} = 0.577 \cdot \frac{3.3^2}{1} = 6.28 \, ohms \qquad (7\text{-}18)$$

However R_{opt} of the other types of PAs needs to be

$$R_{opt} = 0.5 \cdot \frac{V_{DD}^{2}}{P_{out}} = 0.5 \cdot \frac{3.3^{2}}{1} = 5 \, ohms \tag{7-19}$$

Since the matching network or the metal interconnecting lines have a finite resistance, it is very difficult to maintain low output impedance. So, if the required output resistance becomes larger, it is good news for designer!

Second, a class-E power amplifier is suitable for CMOS implementation because class-E power amplifiers use transistors as switches and CMOS devices are excellent switches. Finally, the drain efficiency of class-E PA is greater than for any other PA.

So far, linear class-A, B, C, and AB PAs, and nonlinear class-F and E PAs have been discussed. The class-E PA has a several advantages as shown. Before designing the class-E PA, developing the monolithic inductor model and CAD optimisation are briefly discussed.

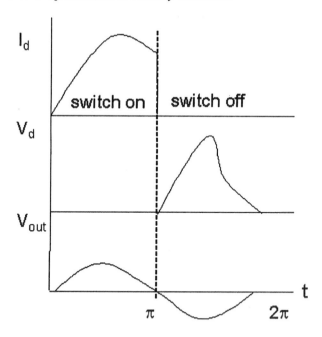

Figure 7-13 Ideal Class-E PA voltage and current waveform

2.   DESIGN OF POWER AMPLIFIER

Typical specifications of a power amplifier are
Output power: 1 Watt (30 dBm)
Drain Efficiency : >50 %
Gain: > 30 dB
Frequency: 900 MHz
Application: GSM (GMSK modulation)
Envelop: constant

Since GSM uses nonlinear digital modulation, GMSK, a nonlinear power amplifier can be used. To obtain 30 dB gain, three stages are needed as shown in Fig. 7-14(b). To achieve an overall 50 % drain efficiency, the last stage drain efficiency is very important as shown in Fig. 7-14(a). Therefore, in this example, a class-AB stage is used for 1st stage and class-E stages are used for 2nd and 3rd stages of the power amplifier. To obtain 1W output power, the resistance seen looking into the matching network to the 50Ω antenna needs to be about 5Ω. Unfortunately, the series resistance of the spiral inductor is greater than 5Ω, and since it forms a voltage divider with the output load, it is impossible to achieve 1W output power with more than a 5Ω parasitic resistance. Hence, the bond wire inductor with its extremely small series resistance is used in the third stage.

To reduce the voltage stresses on the MOSFETs, cascodes are used in the first thru third stages. *HSPICE* simulations were performed using the TSMC 0.35μm CMOS parameters.

3.   OPTIMIZATION OF POWER AMPLIFIER

To optimize this particular RF power amplifier, 22 design parameters need to be optimized where the cost function is aimed at 55% drain efficiency with at least 1.2W output power. The most important cost function components of a switched power amplifier are output power and drain efficiency. The cost function of the power amplifier is

$$Cost = \frac{abs(Pout - 1.2)}{1.2} \cdot weight\_Pout + \frac{EFF - 55}{55} \cdot weight\_EFF$$

(7-20)

According to Eq. (7-20), output power needs to be exactly 1.2 Watt, and drain efficiency needs to be greater than 55 to minimize cost function. In this simulation, weight_Pout and weight_EFF are each 50.

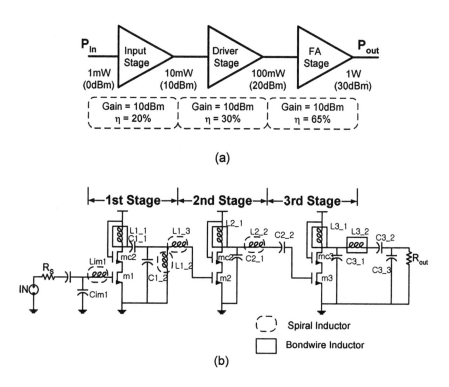

(a)

(b)

Figure 7-14 Three-stage CMOS power amplifier (class-AB, class-E, and class-E) showing gain and drain efficiency (η) distribution

Table 7-1 summarizes parameters that are shown in Fig. 7-14 before optimization and after optimization. As shown in Table 7-1 and Fig. 7-14, Most of the parameters are changed. Especially, RF chokes (L1_1, L2_1, and L3_1) become smaller than ideal case to minimize the series resistance of the RF chokes.

The desirability of parasitic-aware synthesis is vividly illustrated in Fig. 7-15 and table 7-2. With ideal parasitic-free inductors on all three stages, the output power is 1.2W and the drain efficiency is 58% for an input power level of 0dBm. Using parasitic-laden spiral and bond-wire inductors, but before optimization, the output power drops to 0.6W and the drain efficiency decreases to only 30% with 0dBm of input power. Clearly, the parasitics of the passive components significantly degrade the performance of the power amplifier by effectively de-tuning it. Also, Fig. 7-15 shows the simulation results with parasitic-laden inductors after parasitic-aware optimization. Parasitic-aware RF synthesis is essential in the design of efficient high-performance RF circuits that are robust to process, temperature, and voltage variations. As summarized in table 7-2, the use of CAD optimization significantly improves circuit performance; the final design including passive parasitics meets the original design specifications

Table 7-1 Summary of output power, drain efficiency, and gain before optimization and after optimization with power amplifier

Specification	Ideal inductors	Real inductors	
		w/o opt.	w/ opt.
Input Power	0 dBm	0dBm	0dBm
Output Power	1.2Watt	0.6Watt	1.2Watt
Gain (@Pin= 5dBm)	30 dBm	27 dBm	30 dBm
Drain Efficiency	58%	30%	55%

Table 7-2 Summary of the parameters before optimization and after optimization

Value	Ideal	After Opt.	Value	Ideal	After Opt.
Ciml	7.5p	8.6p	C2_1	6.6p	0.68p
Liml	5.5n	6.8n	C2_2	6.6p	0.67p
Bias1	1 V	0.7v	L2_1	10n	3.9n
M1,MC1	2mm	1.25mm	L2_2	9.51n	3.8n
C1_1	3mm	11p	Bias3	1 V	1.3v
C1_2	4mm	0.48p	M3,MC3	15mm	12.4mm
L1_1	5mm	8.7n	C3_1	5.3p	.14p
L1_2	6mm	16n	C3_2	14p	14.1p
L1_3	7mm	2.85	C3_3	14p	7.9p
Bias2	8mm	1 V	L3_1	11.7n	1.54n
M2,MC2	9mm	3.1m	L3_2	5n	4.7n

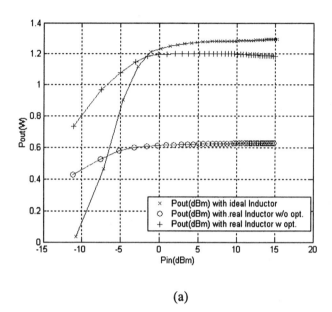

(a)

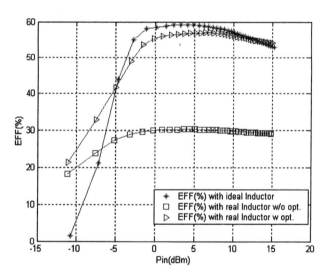

(b)

Figure 7-15 Simulation results with ideal inductors and parasitic-laden inductors before and after optimization

4.   POST PVT OPTIMIZATION

As mentioned in Chapter 3, process, voltage, and temperature variations are very critical in practical circuit designs intended for high-volume manufacturing. However, it takes a long time to simulate the PVT variations. So to save optimization time, post PVT optimization is proposed. The post-PVT simulation m ethodology m eans to run Monte C arlo simulations w ith different temperature, voltage and process parameters, then find the average and standard deviation of the cost function and then re-define the "optimum" solution. So, the optimum solution after PVT variation may be same as or different from the optimum solution before the PVT simulations. To find the optimal solution, the simulator runs the 5 optimization loops, and 38 different solutions are found. Monte Carlo simulations are performed with the 38 different solutions; note that one Monte Carlo simulation requires 340 iterations. The variation of components and voltage are ±10 % and the variation temperature is from 0 to 75 C.

Figure 7-16 shows the a verage and s tandard deviation of output power versus drain efficiency. As shown in Fig. 7-16, even though each individual optimization solution can satisfy the specifications of the PA which are greater than 50% drain efficiency, and 1W power output, not all of the Monte Carlo simulation results satisfy the specifications. Figure 7-17 shows the Monte Carlo simulation results of the (a) best case and (b) worst case of Fig. 7-16. A of Fig. 7-16 represents a best case example and B of Fig. 7-16 represents worst case example. As shown, in the best case, the solutions are distributed in small area; that is, the solution is not sensitive with respect to the PVT variations, so this d esign is robust and easy to fabricate. On t he other hand, in worst case, the solutions are very sensitive, so it is difficult to manufacture due to PVT variations. Therefore, PVT optimization is essential because it determines if the circuit is cost effective in manufacturing.

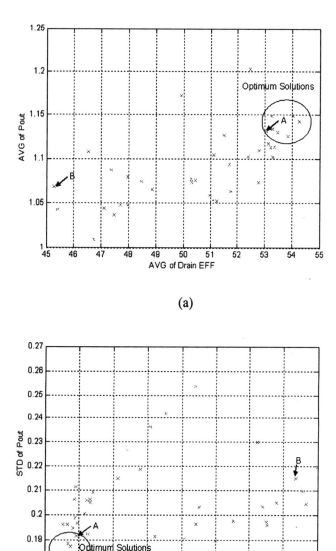

(a)

(b)

Figure 7-16 POST PVT optimization results (a) average of optimum solution (b) Standard
deviation of optimum solution after Monte Carlo Simulation

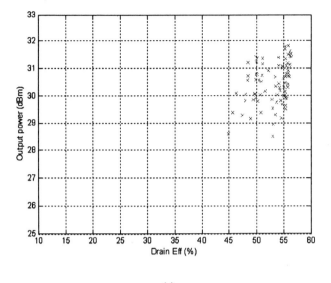

(a)

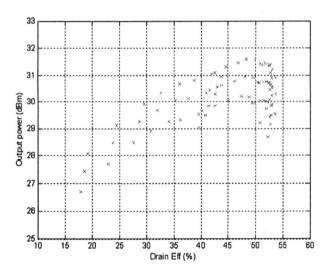

(b)

Figure 7-17 Monte Carlo simulation results of (a) best case(A) and (b) worst case(B) of Fig. 7-16

REFERENCES

[1] T. Lee, *The Design of CMOS Radio-Frequency Integrated Circuit*, Cambridge University Press, 1998.

[2] S. Cripps RF *Power Amplifiers for Wireless Communications*, Artech House Publishers 1998.

[3] N.O. Sokal and A.D. Sokal, "Class-E: A new class of high-efficiency tuned single-ended switching power amplifiers," *IEEE J. Solid-State Circuits*, pp. 168-176, June 1975.

[4] F.H. Raab, "Idealized operation of the class-E tuned power amplifier," *IEEE Trans. on Circuits and Systems*, pp. 725-735, Dec. 1977

[5] K. Tsai, P.R. Gray, "A 1.9 GHz 1-W CMOS Class-E Power Amplifier for Wireless Communications," *IEEE Journal of Solid State Circuits*, pp.962-970 July 1999

[6] T. Sowlati, A.C.T. Salama, J. Sitch, G. Rabjohn, and D. Smith, "Low voltage, high efficiency GaAs Class-E power amplifiers for wireless transmitters," *IEEE J. Solid-State Circuits*, vol. 30, pp. 1074-1079, Oct. 1995.

[7] G.K. Wong and S.I. Long, "An 800 MHz HBT class-E amplifier with 74% PAE at 3.0 volts for GMSK," *GaAs IC Symposium*, pp. 299-302, 1999

[8] K. Choi, D. Allstot, and S. Kiaei "Parasitic-Aware Synthesis of RF CMOS Switching Power Amplifiers," *International symposium on circuit and system*, pp. 269 -272 May 2002

Chapter 8

OPTIMIZATION OF ULTRA-WIDEBAND AMPLIFIERS

1. CMOS ULTRA-WIDEBAND AMPLIFIERS

Wideband amplifiers are widely used in many radio frequency (RF) and high-data rate communication systems such as satellite transceivers, pulsed radar, optical receivers, etc [1]. For such applications, a distributed amplifier topology is often used because it overcomes the classical gain-bandwidth tradeoff of amplifiers. However, distributed amplifiers have not been integrated in Si-CMOS process until recently because of the relatively low cutoff frequency (f_t) of CMOS process and its poor on-chip passive components compared to GaAs and SiGe counterparts. The former problem is somewhat solved due to its aggressive scaling capability originated from the advance of digital integrated circuit technologies. On the other hand, the latter is still the bottleneck of design because the performance of Si-CMOS on-chip passive components is not scaled unlike that of active components. In this chapter, we will study CMOS distributed amplifier design using parasitic-aware design technique.

1.1 Distributed amplification theory

In general, the gain and the bandwidth of an amplifier cannot be simultaneously increased beyond a certain limit that is determined by the process technology used [2]. This is known as a classical gain-bandwidth tradeoff as shown in Fig. 8-1. This plot shows a frequency response of an amplifier in CMOS. In order to increase the bandwidth of this amplifier, it is the only way to trade the gain of the amplifier for the bandwidth since the size of transistors should be decreased for the bandwidth for a given bias and this in turn will reduce the gain of the amplifier by the same amount as the increased bandwidth.

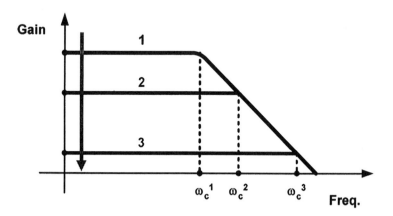

Figure 8-1 Classical gain-bandwidth product limitation

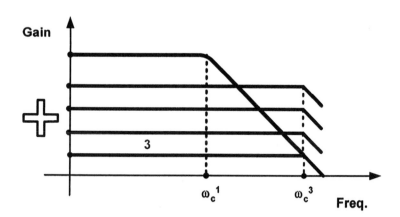

Figure 8-2 Concept of an additive distributed amplification

Conceptually, this limitation can be overcome by adding many stages with gain-bandwidth characteristic a s the third case in the plot (Fig. 8-2). This is achieved by separating parasitic capacitances of transistors using inductors and combining the gain of transistors in additive fashion. This will construct an L-C ladder circuit as shown in Fig. 8-3.

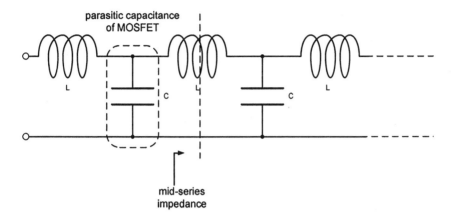

Figure 8-3 L-C artificial transmission line

This L-C ladder circuit is also called *artificial transmission line* because it resembles transmission line characteristics such as impedance and phase. In fact, if the artificial transmission line has an infinite number of sections with infinitesimal inductance and capacitance, it converges to the ideal transmission line without any dissipation and dispersion. However, it shows discrepancy from the ideal transmission line in characteristic impedance and bandwidth since it is impossible to make an infinite L-C ladder in practice. Thus, the artificial transmission line should be terminated and the number of sections decides the quality of the artificial transmission line. Figure 8-4 shows the input impedance with respect to the number of sections. As can be seen in this figure, t he a rtificial transmission line with a s mall number of sections shows a very poor frequency response with a relatively narrow bandwidth. In addition, the magnitude of ripples in passband is inversely proportional to the number of sections, and the noise figure of a distributed amplifier improves with the number of sections [2],[3]. Therefore, it is necessary to use as many numbers of sections as possible for broad bandwidth.

The mid-series image impedance in this case is

$$Z = \sqrt{\frac{L}{C}(1 - \omega/\omega_c^2)} \qquad\qquad (8\text{-}1)$$

$$\omega_c = \frac{2}{\sqrt{LC}} \qquad\qquad (8\text{-}2)$$

where ω_c is the cutoff frequency of the artificial transmission line [2]. Note that the cutoff frequency is not same as the resonant frequency of LC tanks. This cutoff frequency decides the bandwidth of the distributed amplifier.

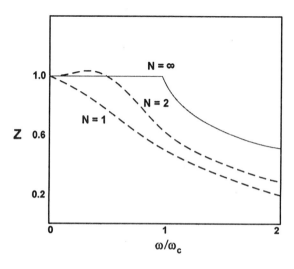

Figure 8-4 Normalized impedance vs. normalized frequency with the number of sections

1.2   CMOS distributed amplifier

In a distributed amplifier, transistors are connected in parallel with the gates and drains of transistors separated by inductors, thus constructing two artificial transmission lines at input and output as shown in Fig. 8-5. The input signal propagates down the input artificial transmission line, and a current signal is generated in the output line by the g_m generator of each transistor encountered. If proper phase synchronization is achieved for both input and output lines, the output signals propagating to the right will be constructively combined at the output in this figure. However, the output signals in the reverse direction should be absorbed in the termination resistor to prevent reflection since these signals are not synchronized. These reverse signals are not only too small to be used for amplification but also strongly dependent o n t he frequency, t hus d egrading the f requency response o f t he amplifier. If the phases of input and output are assumed to be matched, the overall gain of a distributed amplifier in the passband will be proportional to the number of sections, the transconductance of transistors, and the output impedance of the amplifier approximately as

$$|A_V| \cong \frac{N \cdot g_m \cdot Z_o}{2} \qquad (8\text{-}3)$$

where N is the number of sections and Z_0 is the output impedance. Note that a factor of 2 is introduced in the denominator to account for the current in the reverse direction.

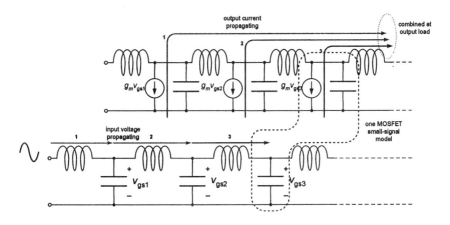

Figure 8-5 Small-signal model of a CMOS distributed amplifier

1.3   Effects of loss in CMOS distributed amplifiers

In theory an ideal distributed amplifier has no limit on the number of sections t hat could b e connected t ogether [4]. S ince the g ain i s i ncreasing with the number of sections, this in essence gives it a potentially infinite gain-bandwidth product. In reality, however, practical limitations put a limit on the maximum number of sections in series that can be used. The limitation is due to the attenuation that is present in the artificial transmission lines of the amplifier such as metal line resistances, finite Q of inductors, drain-to-source small-signal resistances, the gate resistance, etc.

The attenuation decreases the strength of the signals propagating through the lines, and eventually cancels out the amplification effect of active devices. Therefore there are an optimum number of sections which will maximize the gain, given by

$$N_{OPT} = \frac{\ln(A_d / A_g)}{A_d - A_g} \tag{8-4}$$

where A_g and A_d are the attenuation per section of the artificial transmission line with the gate and the drain connected, respectively [5]. The attenuation factor can be derived from the propagation function for a section of the line. For the artificial transmission lines in Fig. 8-6, the propagation function is given by

$$\theta_g = A_g + j\phi_g = \mathrm{acosh}\left(1 + \frac{1}{2} \cdot \frac{(R_{lg} + j\omega L_g)}{(R_g + 1/j\omega C_{gs})}\right) \tag{8-5}$$

$$\theta_d = A_d + j\phi_d = \mathrm{acosh}\left(1 + \frac{(R_{ld} + j\omega L_d)(G_{ds} + j\omega C_d)}{2}\right) \tag{8-6}$$

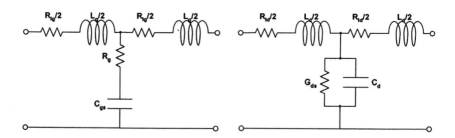

Figure 8-6 Losses associated with input line and output line

There are mainly three different losses in the artificial transmission lines, namely, the gate resistance R_g, the drain-to-source resistance R_{ds}, and the inductor series resistance R_{lg}.

The gate resistance R_g consists of two components: the distributed gate electrode resistance (R_{gel}) and the distributed channel resistance seen from the gate (R_{gch}). The gate electrode resistance can be estimated with the following formula for an interdigitated gate structure connected on both sides of the gate [6]

$$R_{gel} = \frac{R_{sh}W/L}{12N^2}$$ (8-7)

where R_{sh} is the sheet resistance of the gate, and N is the number of fingers. The distributed channel resistance is estimated as

$$\frac{1}{R_{gch}} = \gamma \left(\frac{1}{\int dV/I_d} + \frac{1}{qL/\eta W\mu C_{ox}kT} \right)$$ (8-8)

where γ is a parameter accounting for the distributed nature of the channel resistance and C_{ox}. γ is equal to 12 if the resistance is uniformly distributed along the channel. The gate resistance is very critical for this type of applications since even a small value of the gate resistance becomes comparable to the impedance of the gate capacitance at very high frequency and hence v_{gs} can be much less than the input voltage as shown in Fig. 8-7, which affects directly the overall gain of the amplifier.

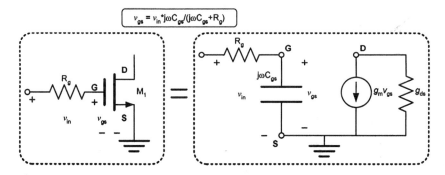

Figure 8-7 Effects of gate resistance at high frequency

Another difference from the ideal characteristics of transmission lines is the shunt resistance r_{ds} or g_{ds} in the output line other than pure reactive components. This works as another loss mechanism by providing undesirable resistive signal paths. There are several techniques reported in literatures. One solution is adding a negative conductance to cancel out g_{ds} in the output line [7]. However, this technique should be applied very carefully;

otherwise it may make the amplifier unstable. Another method is increasing the output impedance of active devices using cascode topology [8]. This technique not only reduces the loss associated with the active device, but also increases the reverse i solation between the input and the output. This increases the number of transistors and may need additional bias circuitries.

The t hird l oss originates from o n-chip inductors. F igure 8-8 s hows t he typical on-chip spiral inductors and the equivalent circuit parasitic model. In addition to the resistive loss from the finite metal conductance, the capacitive loss to the lossy silicon substrate degrades the performance of on-chip inductors significantly.

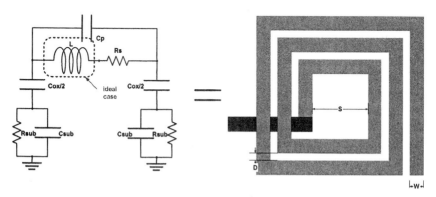

Figure 8-8 Parasitic model of CMOS on-chip spiral inductors

2.   DESIGN OF CMOS ULTRA-WIDEBAND AMPLIFIER

To understand the device characteristic and explore the feasibility of 0.35-μm CMOS technology for this broadband application, the cutoff frequency (f_t) of a 2-port active device is estimated where the current gain is unity. Figure 8-9 shows that f_t is wide enough to be used for the 8GHz amplifier.

For each amplification cell, a cascode structure is used to reduce the Miller effect by imposing low impedance at the drains of the amplifying devices. Cascoding also reduces capacitive coupling between the artificial transmission lines and increases gain flatness, reverse isolation, stability, and input and output impedance matching accuracy. The high output impedance of the cascode configuration also reduces the loss a ssociated with amplifier loading on the drain line [9].

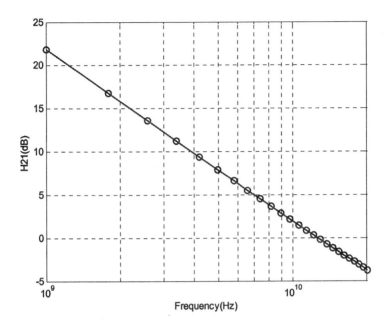

Figure 8-9 H_{21}(dB) against frequency for estimation of f_t

Even though the number of sections in the artificial transmission line is increased, the overall frequency response near the cutoff frequency is degraded by the impedance mismatch between the terminating resistor and the artificial transmission line. Thus, in order to improve the impedance match, a buffer stage called *m*-derived half section is inserted between the terminating resistors and artificial transmission lines. A feature of an *m*-derived half section is that it exhibits identical mid-series image impedance as that of the artificial transmission line. The mid-shunt image impedance of the *m*-derived half section is given as

$$Z_{i\pi} = \sqrt{\frac{L}{C}} \frac{1 - \omega^2 / \omega_o^2}{\sqrt{1 - \omega^2 / \omega_c^2}} \qquad (8\text{-}9)$$

where $\omega_o = \omega_c / \sqrt{1 - m^2}$. This feature can be exploited to provide a fairly constant impedance up to frequencies near cutoff. It has been shown that using *m* values near 0.6 provides near optimal results [10][11]. This technique also reduces a large undesirable peak in the gain response near the cutoff frequency and improves the gain-flatness of the amplifier.

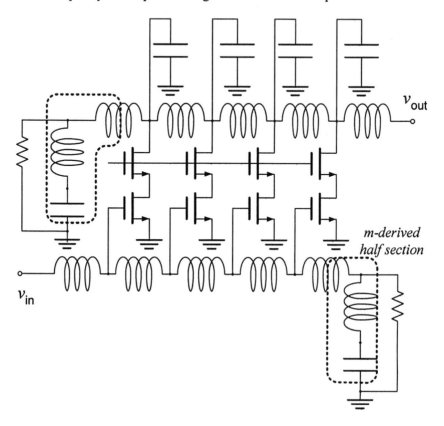

Figure 8-10 Cascode 4-stage distributed amplifier with *m*-derived half section

Using *m*-derived half sections for matching to input and output 50Ω, the complete schematic o f 4-stage d istributed a mplifier i s s hown i n F ig. 8 -10. Note that extra capacitors are added to the output line to make the correct phase match. Running an *HSPICE* simulation of the circuit produced the response shown in Fig. 8-11. The resultant forward gain is about 8dB with ±1.0dB flatness over 8GHz of bandwidth. The reverse isolation (S_{12}) is less than -65dB in the passband due to the cascode device. However, after ideal components are replaced by realistic models, a drastic change in performance is evident. This is demonstrated in Fig. 8-12 and Fig. 8-13, which shows the gain and phase response for the circuit once parasitic-laden inductor models are used, and gate losses are accounted for. As can be seen in this figure, there is a significant drop in gain between the ideal and realistic response. One reason of this degradation is that the large parasitic capacitance associated with the monolithic inductors has offset the line capacitance to the extent that the artificial transmission lines no longer have the correct phase synchronization. As such, the device output currents do not add in phase and therefore do not produce significant gain. In addition, the series loss of the parasitic-laden inductors and parasitic gate resistance directly attenuate the propagating signals, thereby causing a direct drop.

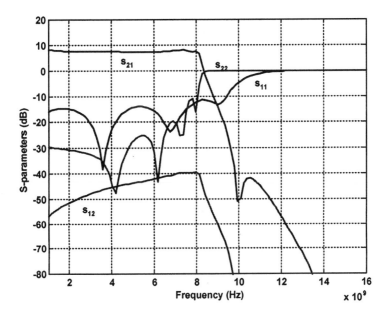

Figure 8-11 S-parameter results for 4-stage distributed amplifier

3. OPTIMIZATION OF CMOS ULTRA-WIDEBAND AMPLIFIER

For a given 0.35-μm CMOS process, the goal is to design a CMOS distributed amplifier with 8dB forward gain (s_{21}) flat over 8GHz. The phase response should be as linear as possible over the passband, and the input and output are to be matched to 50Ω.

Basically, the phase linearity over the wide frequency range is the key of success for a broadband response to combine the propagating signals constructively. Once parasitics are considered, however, the phase match becomes no longer valid. In fact it is virtually impossible to regain the phase match manually owing to the frequency-dependent nonlinear behavior of parasitics. In this section, the gain and phase linearity of a CMOS distributed amplifier are optimized using parasitic-aware design technique.

3.1   Optimization of a CMOS distributed amplifier

Due to the non-ideal effects of realistic design, the order of the problem complexity has increased significantly. To overcome the performance degradation caused b y parasitics, a n efficient c omputer a ided o ptimization should be used in a CMOS distributed amplifier.

The goal of this optimization problem is to maximize the forward gain (S_{21}), the forward gain flatness, and phase linearity. Figure 8-14 shows a single optimization loop and it repeats until it reaches the predefined maximum number of iterations. Initially, *HSPICE* S-parameter ac simulation is performed with the synthesized *netlist*. This simulation run has been designed to result in the average gain, gain flatness in terms of its minimum and maximum value, and phase linearity onto a single output file for an easy file manipulation. Then these results are simply mapped onto a scalar cost function for the evaluation of the synthesized circuit.

3.2   Optimization results

There is a clear difference between the un-optimized and the synthesized distributed amplifiers. The forward gain has been improved by 2dB and the flatness from ±2dB to ±1dB. Note that this improvement has been achieved by regaining the phase match as shown in Fig. 8-12 and 8-13. Loss compensation and improved phase linearity also drastically improve the gain flatness.

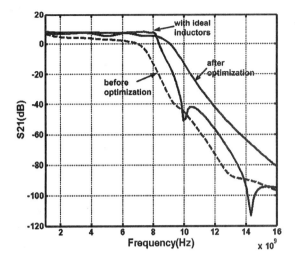

Figure 8-12 Forward gain (S_{21}(dB)) magnitude results

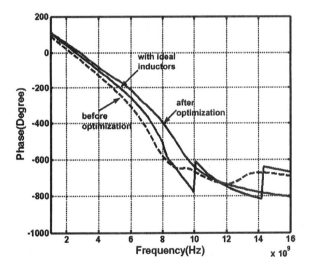

Figure 8-13 Forward gain phase results

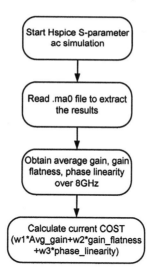

Figure 8-14 Average gain, flatness, and phase linearity optimization loop

REFERENCES

[1] H.T. Ahn and D. Allstot, "A 0.5-8.5GHz fully differential CMOS distributed amplifier," *IEEE J. Solid-State Circuits,* vol. 37, pp. 985-993, Aug. 2002.

[2] T.T.Y. Wong, *Fundamentals of distributed amplification*, Artech House, Inc., Norwood, MA, 1993.

[3] C.S. Aitchison, "The intrinsic noise figure of the MESFET distributed amplifier," *IEEE Trans. Microwave Theory Tech.,* vol. MTT-33, pp. 460-465, June 1985.

[4] E.L. Ginzton, et al., "Distributed amplification," Proc. IRE, vol. 36, pp. 956-969, Aug. 1948.

[5] J.B. Beyer, et al., "MESFET distributed amplifier design guidelines," IEEE Trans. Microwave Theory Tech., vol. MTT-32, pp. 268-275, Mar. 1984.

[6] X. Jin, et al., "An effective gate resistance model for CMOS RF and noise modeling," *IEDM*, pp. 961-964, 1998.

[7] S. Kimura, Y. Imai, Y. Umeda, and T. Enoiki, "Loss-compensated distributed baseband amplifier IC's for optical transmission systems," *IEEE Trans. Microwave Theory Tech.,* vol. MTT-44, pp. 1688-1693-465, Oct. 1996.

[8] S. Deibele and J. B. Beyer, "Attenuation compensation in distributed amplifier design," *IEEE Trans. Microwave Theory Tech.,* vol. MTT-37, pp. 1425-1433, Sept. 1989.

[9] J. H. Park, K. Y. Choi, and D. J. Allstot, "Parasitic-aware design and optimization of a fully integrated CMOS wideband amplifier," *to be published at ASP-DAC*, 2003.

[10] D.M. Pozar, *Microwave Engineering*. Reading, MA: Addison-Wesley, 1990

[11] B. Ballweber, B.M, R. Gupta, and D. Allstot "A fully integrated 0.5-5.5 GHz CMOS distributed amplifier," " *IEEE J. Solid-State Circuits,* vol. 35, pp. 231-239 Feb. 2000

Index